# Development for Sustainable Agriculture

# Development for Sustainable Agriculture

## The Brazilian Cerrado

Edited by

Akio Hosono
*Senior Research Adviser, Japan International Cooperation Agency
Research Institute (JICA-RI), Japan*

Carlos Magno Campos da Rocha
*Former President, Brazilian Agricultural Research Corporation (EMBRAPA), Brazil*

and

Yutaka Hongo
*Former Senior Adviser, Japan International Cooperation Agency (JICA), Japan*

First published 2016 by
PALGRAVE MACMILLAN

Palgrave Macmillan in the UK is an imprint of Macmillan Publishers Limited, registered in England, company number 785998, of Houndmills, Basingstoke, Hampshire RG21 6XS

Palgrave Macmillan in the US is a division of St Martin's Press LLC, 175 Fifth Avenue, New York, NY 10010.

Palgrave is the global academic imprint of the above companies and has companies and representatives throughout the world.

Palgrave® and Macmillan® are registered trademarks in the United States, the United Kingdom, Europe and other countries

ISBN: 978–1–137–43134–9

This book is printed on paper suitable for recycling and made from fully managed and sustained forest sources. Logging, pulping and manufacturing processes are expected to conform to the environmental regulations of the country of origin.

A catalogue record for this book is available from the British Library.

Library of Congress Cataloging-in-Publication Data

Development for sustainable agriculture: the Brazilian cerrado/[edited by] Akio Hosono (senior research adviser, Japan International Cooperation Agency Research Institute (JICA-RI), Japan), Carlos Magno Campos da Rocha (former president, Brazilian Agricultural Research Corporation (EMBRAPA), Brazil), Yutaka Hongo (former senior adviser, Japan International Cooperation Agency (JICA), Japan).
    pages cm
    Includes bibliographical references.
    ISBN 978–1–137–43134–9 (hardback)
    1. Sustainable agriculture – Brazil. 2. Sustainable development – Brazil. 3. Cerrados – Brazil. 4. Grain – Brazil. 5. Agricultural productivity – Brazil. 6. Poor – Employment – Brazil. 7. Food security – Brazil. 8. Environmental protection – Brazil. 9. Brazil – Economic conditions. 10. Brazil – Environmental conditions. I. Hosono, Akio. II. Magno Campos da Rocha, Carlos, 1972– III. Hongo, Yutaka, 1948–
  S475.B7D48 2015
  631.50981—dc23

                             2014049567

# Contents

## Part I    Development of Cerrado Agriculture

# List of Figures

# List of Tables

# Foreword

As the world's population is expected to approach nine billion by the end of this century, with most of this population growth likely to take place in the developing world, the international community is now faced with the pressing need to address global food security, particularly in developing countries, while at the same time conserving the environment. Brazil is a remarkable pioneer in this field, achieving food production and environmental protection simultaneously.

Brazil achieved an epoch-making breakthrough to become a net exporter of grain by converting barren land into one of the most productive agricultural areas in the world. Starting in the mid-1970s, the tropical savanna area called the Cerrado was transformed into one of the world's largest grain-growing regions. This realization of modern, upland, rain-fed grain production in a tropical region happened in just a quarter of a century, a transformation described as "one of the great achievements of agricultural science in the 20th century,"[1] by Nobel Prize laureate Norman E. Borlaug.

The Cerrado development has had a crucial impact on Brazil's national economy, enabling impressive poverty reduction through job generation, socially inclusive growth, the improvement of nutrition, and increases in food supplies. For environmental and ecological conservation, many innovative technologies and institutions have been developed and introduced. The experience of the Cerrado development could be a valuable lesson for other developing countries working toward nutritional and food security, value chain development, job creation, socially inclusive growth, and sustainable development. Brazil and Japan engaged in technical and financial cooperation programs in this region, engaging both public and private sectors to increase grain production in the Cerrado and contribute to the world food supply. Through demonstration and diffusion effects, the cooperation decisively contributed to the increase in agricultural productivity and the socioeconomic development of the Cerrado region.

This book is the outcome of a research program conducted by the Japan International Cooperation Agency Research Institute with the participation of Brazilian and Japanese researchers and professionals, many of whom played key roles in the cooperation programs. I am convinced that this book will offer lessons in the socially inclusive and sustainable development of tropical agriculture and agro-industrial value chains, and hence in the attainment of global food security.

Akihiko Tanaka
*President, Japan International Cooperation Agency*

## Note

1. httl://www.worldfoodprize.org/laureates/Past/2006.htm

# Foreword

This book is the result of comprehensive research on the development of Cerrado agriculture, evaluating the reasons for its success and the contribution it has made to a stable food supply and regional development. As one of those who have been involved in promoting Cerrado agriculture for over 40 years, I sincerely welcome this book being published.

My involvement with Cerrado development goes back to 1971, when I was appointed State Secretary of Agriculture in Minas Gerais by Governor Rondon Pacheco. I was then a professor at the Federal University of Lavras, researching the possibility of Cerrado agriculture. At that time, 65 per cent of Minas Gerais State was covered by Cerrado vegetation, which was of almost no use except as fodder for cattle. The first step was the Alto Paranaíba Agricultural Development Program (PADAP), in cooperation with the Cotia Agricultural Cooperative (CAC). Farmers in other states who were interested in experimentation were invited to participate. It was a great success and became the starting point of Cerrado agricultural development.

After taking over as Minister of Agriculture in the federal government under President Geisel in 1974, I undertook the Cerrado Development Program (POLOCENTRO) the following year. I was involved in the planning of the Japanese–Brazilian Cooperation Program for Cerrados Development (PRODECER), implemented in 1979. I also helped to establish the Cerrados Agricultural Research Center (CPAC) under the Brazilian Agricultural Research Corporation (EMBRAPA) in 1975 to act as a main research institution for the Cerrado region. Meanwhile, many relevant agencies were engaged in efforts to conserve the environment. The combination of modern agricultural production technologies and sound environmental practices has transformed the "arid savanna" into productive and sustainable agricultural production systems feeding poor people in Brazil and elsewhere in the world.

The Cerrado is now fertile and is one of the largest agricultural regions in the world. The soil has changed drastically since the 1970s. In 2006, Edson Lobato, one of the CPAC's research soil fertility leaders, and I were honored to become the first Brazilian recipients of the World Food Prize for our contribution to the Cerrado's agricultural development – international recognition of the Brazilian contribution to tropical agriculture around the world.

This book describes Cerrado agriculture from a variety of viewpoints over a long span of time, with contributors from Japan and Brazil. From the Brazilian side, chapters have been contributed by researchers and officials who played key roles in agencies providing support to Cerrado agriculture, such as EMBRAPA and CPAC, the Agricultural Promotion Company (CAMPO), PRODECER and the Brazilian Institute for Environment and Renewable Natural Resources (IBAMA), the center of environmental administration in Brazil. From the Japanese side, chapters have been contributed by researchers from the Japan International Cooperation Agency (JICA), which served as an administrative agency for providing Japanese cooperation in Cerrado agriculture.

The recent history of Cerrado agriculture provides a wealth of experiences that will repay analysis by those interested in the development of tropical agriculture. There are significant lessons to be learned in devising and executing well balanced, socially inclusive regional development, as well as in managing agricultural and environmental protection, in addition to food security. The book is a significant intellectual contribution in providing information and insights not to be found elsewhere. I hope that many people who are interested in global food issues, sustainable tropical agriculture, and socially inclusive agricultural development will benefit from it.

Alysson Paulinelli
*Former Minister of Agriculture, Brazil*

# Acknowledgements

The editors are most grateful to all the staff of the JICA Research Institute and the JICA office in Brazil, who provided efficient logistic support to us in the preparation of this volume. The concept originated at a seminar organized as a side-event under the auspices of JICA and the Brazilian Cooperation Agency (ABC) on the occasion of the Rio+20 Conference in Rio de Janeiro in 2012, in which we participated as presenters. We would especially like to thank Dr. Akihiko Tanaka, Former President of JICA, and Hiroshi Kato, Vice-President of JICA and former Director of the JICA Research Institute for strongly encouraging us to prepare this volume.

Several other participants in this event are among the contributors to this volume. We would like to thank them and the institutions to which they belong, especially the Brazilian Agricultural Research Corporation (EMBRAPA), Brazilian Institute for Environment and Renewable Natural Resources (IBAMA), EMBRAPA's Cerrado Agricultural Research Center (CPAC), and the Company of Agricultural Promotion (CAMPO).

This volume has also benefited from the input of the many key people who kindly accepted our invitations to be interviewed and gave us invaluable information on Cerrado development: Tomomi Ashikaga, former official of the Ministry of Agriculture, Forestry and Fisheries (Japan) and former Executive Director of the Japan–Brazil Agricultural Development Corporation (JADECO), Ishidoro Yamanaka, former aide to the Minister of Agriculture, Livestock and Supply (Brazil), Dr. Fumio Iwata, former Executive Researcher of the Japan International Research Center for Agricultural Sciences (JIRCAS), Dr. Osamu Ito, Dr. Yasuhiro Tsujimoto, and Dr. Tetsu Tobita (JIRCAS), Yoshihiko Horino, former researcher of the National Center for Vegetable Research (CNPH), Dr. Wenceslau J. Goedert, former director general of the Cerrado Agricultural Research Center (CPAC), Dr. Elmar Wagner, former Director General of CPAC, Dr. Claudio Alberto Bento Franz, former Director General of CPAC, Dr. Sebastião Pedro da Silva Neto, Dr. Leide Rovênia Miranda de Andrade, and Dr. Plínio Itamar de Melo de Souza (CPAC), Vasco Praca Filho, Mayor of Paracatu County (MG), Wander Cordeiro, Gentaro Arimura, Autair Pereira de Camargo, Washington Hiroyuki Endo, Alberto Mendes Costa, Ademar Sagae, and Masao Yamamoto (Paracatu, MG), Norton Komagome (Entre Ribeiros, Paracatu, MG), Emiliano Pereira Botelho, President, CAMPO, Mitsutoshi Akimoto, Vice-President, CAMPO,

Dr. Paulo Afonso Romano, former President, CAMPO, Dr. Aluízio Fantini Varério, former President, Rural Minas, and former Director of Technology, CAMPO, Marcio Xavier Bartels, former researcher, CAMPO Paracatu Office, Alvaro Luiz Orioli (CAMPO), Horácio Takeu Muraoka, Executive Director, COOPADAP (Sao Gotardo, MG), Nicolau Minami, Executive Director, COOPADAP (Sao Gotardo, MG), Jorge Fukuda and Tamio Sekita (Sao Gotardo, MG), Dr. Alysson Paulinelli, former Minister of Agriculture, Livestock and Supply, Adilson Heidi Sujuki, Executive Director, COOBAHIA, Jorge Tadashi Koyama, Executive Director, COOPROESTE, Walter Yukio Horita, President, Horita Empreendimento Agricola (Barreiras, BA), Sérgio Setsuo Sato (Barreiras, BA), Dr. Pedro António Arrães Pereira, President EMBRAPA, Dr. Eliseu Roberto de Andrade Alves, former President, EMBRAPA, Dr. Kazumitsu Matsumoto, CENARGEM/EMBRAPA (Brasilia), Dr. José Roberto Rodrigues Peres, Secretary General, EMBRAPA and Director General CPAC, Ricardo Villela de Souza, SEBRAE (Brasilia), Marco Farani, General Director of ABC, Dr. Joao Roberto Rodriguez, former Minister of Agriculture, Livestock and Supply, Americo Utsumi, former Vice-President of Cooperative Cotia, Carlos Otsubo, former executive of Cooperative Cotia, José Wilson Siqueira Campos, Governor, State of Tocantins, Ricardo Benedito Khouri, Executive Director COAPA (Pedro Afonso, TO), Edú Laudi Paskoski, Secretary of Agriculture and Environment, Lucas do Rio Verde, Fábio Ricardo Raabe, Department of Agricultural Development, Lucas do Rio Verde, Shoiti Minami (Lucas do Rio Verde), Marino José Franz, Mayor, Lucas do Rio Verde, and Siro Iida (Piuva, MT), José Júlio Eduardo Chagas, Mayor of Pedro Afonso, Joci Piccini, Vice-Mayor, and Lucas do Rio Verde (MT), Ségio Masao Murakami (Lucas do Rio Verde), and Dr. Edson Eyji Sano (IBAMA).

We would like to acknowledge Professor Tetsuo Mizobe of Nihon University and Professor Shoichi Ito of Kyushu University for writing the background papers for this volume. We are thankful to Keisuke Takahashi and Gustavo H. Meireles for their support in preparing and compiling information on Cerrado agriculture. We owe a lot to Takahashi and Meireles, who made a great contribution by transcripting recorded interviews of around 75 hours.

We would like to express our sincere gratitude to the city of Lucas do Rio Verde for providing us with detailed information on the development of the city and permission to use the photograph of the city that appears on the front cover of this book. We are grateful to Dr. Keiichi Tsunekawa and Dr. Naohiro Kitano for their continuous support for our study. We are thankful to Yasuhiko Sato, Chief Editor of the JICA

Research Institute, together with Koji Yamada, Haruko Kamei, Kumiko Kasahara, Maiko Takeuchi, and Aiko Oki for providing professional assistance during the preparation of this volume.

In addition, Eliseu Alves would like to thank the following for their assistance in the preparation of Chapter 6: Antônio Jorge de Oliveira, Cláudio Costa Cardoso, Geraldo da Silva e Souza, and Renner Marra.

Akio Hosono, Carlos Magno, Yutaka Hongo

# Note on Cover Image

The city of Lucas do Rio Verde in the Cerrado region, as it appears today. A vast crop field (light green area in the photograph) surrounds the city, stretching as far as the horizon. Alongside the river running across the center of the city are environmental conservation zones, where virgin nature is preserved (dark green area). The city itself consists of industrial zones with agro-processing plants, commercial districts, and residential areas, including modern houses for new arrivals.

The city of Lucas do Rio Verde started as one of the focal points of the Japanese–Brazilian Cooperation Program for Cerrados Development (PRODECER) in the State of Mato Grosso. It was ranked 8th in 2011 among the 5,564 municipalities of Brazil according to the socio-economic development criteria of FIRJAN (Federation of Industries of the State of Rio de Janeiro). In 2006, the city was voted the Most Environment Friendly Municipality in the country by the *Jornal do Brasil*, one of the most popular national newspapers.

# Map of Cerrado

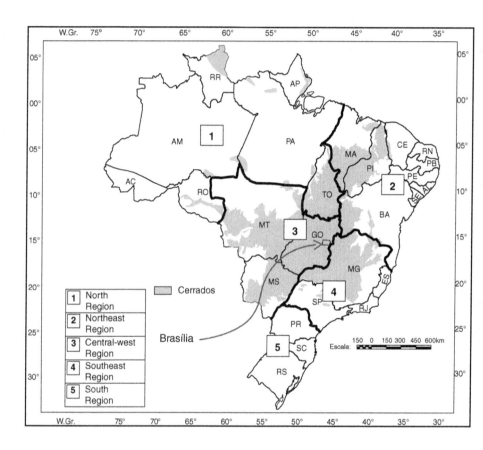

This map is based on the one used by the Cerrado Agricultural Research Center (CPAC) of the Brazilian Agricultural Research Corporation (EMBRAPA) since its foundation in 1975. In 2004, the Brazilian Institute of Geography and Statistics (IBGE) published a map of Brazilian biomes,[1] in which the Cerrado biome (ecosystem), as identified, does not exactly overlap with the area on the CPAC map. While the CPAC map includes many detached zones of the Cerrado biome in the midst of other biomes, the IBGE map has no such areas. The differences arise from difficulties in specifying the precise boundaries of the Cerrado because the transition to other ecosystems is gradual. Because of these differences, estimates of the size of Cerrado vary. For example, Adámoli et al.[2] estimated it to be 203.7 million ha, while Sano et al.[3] estimated it at 207.4 million ha, both using the CPAC map. The IBGE estimated 204.7 million ha using its own map.[4] Some statistics, such as agricultural production, could also differ depending on the map that is used. Therefore, special care is advised in order to avoid any confusion.

As the IBGE map was published in 2004 and the CPAC map had been used as the basic reference of the Cerrado region for about three decades, any analysis of the historical evolution of the Cerrado agriculture has to use the CPAC map, as Part I of this book does, while in Part II, both maps are used.

## Notes

1. http://www.ibge.gov.br/home/presidencia/noticias/21052004biomashtml.shtm. For details of the differences between the CPAC's and IBGE's maps, see Chapter 8.
2. Adámoli, J., Macedo, J., Azevedo, L.G., and Madeira Netto, J. (1985) "Caracterização da região dos Cerrados" in W. Goedert (ed.), *Solos dos Cerrados. Tecnologias e Estratégias de Manejo.* São Paulo: Nobel; Brasília: EMBRAPA Centro de Pesquisa Agropecuária dos Cerrados, pp. 33–74.
3. Sano, E.E., Barcellos, A.O., and Bezerra, H.S. (2000) Assessing the spatial distribution of cultivated pastures in the Brazilian savanna. *Pasturas Tropicales,* 22(3), pp. 2–15.
4. See the above web page.

# Notes on Contributors

**Eliseu Alves** is an EMBRAPA researcher and a former president of EMBRAPA and the Companhia de Desenvolvimento do Vale do São Francisco e do Parnaíba (CODEVASF).

**Emiliano Pereira Botelho** is President of CAMPO.

**Carlos Magno Campos da Rocha**, who has an MSc in Animal Nutrition, is an agronomist and was president of EMBRAPA from 1989 to 1990, and director general of CPAC from 1988 to 1989, and from 1996 to 2003.

**Wenceslau J. Goedert**, who has a PhD in Soil Fertility, is an agronomist and was director general of CPAC from 1986 to 1988, and from January to July 2011.

**Yutaka Hongo** is a former senior adviser for the Japan International Cooperation Agency (JICA).

**Akio Hosono**, who has a doctorate in Economics, is senior research adviser and a former director of the Japan International Cooperation Agency Research Institute (JICA-RI).

**Roberto Rodrigues** is a former minister of Agriculture of Brazil.

**Edson Eyji Sano** is a senior remote sensing researcher at the Brazilian Agricultural Research Corporation (EMBRAPA) under the Ministry of Agriculture and head of the Remote Sensing Center of the Brazilian Institute for Environment and Renewable Natural Resources (IBAMA) under the Ministry of Environment.

**Elmar Wagner**, who has an MSc in Hydrology, is an agronomist and was director general of CPAC from April 1976 to August 1984.

# List of Abbreviations

| | |
|---|---|
| ABC | Brazilian Cooperation Agency |
| BA | Bahia (Brazilian state) |
| CAC | Cotia Agricultural Cooperative |
| CAMPO | Agricultural Promotion Company |
| CNPH | National Center for Vegetable Research |
| CPAC | Cerrado Agricultural Research Center |
| EBDA | Empresa Baiana de Desenvolvimento Agricola |
| EMBRAPA | Brazilian Agricultural Research Corporation |
| EMBRATER | (Empresa Brasileira de Asistencia Tecnica e Extensão Rural) Brazilian Agency for Rural Extension and Technical Assistance |
| EPAMIG | (Empresa de Pesquisa Agropecuária de Minas Gerais) Agricultural Research Company of Minas Gerais State |
| FIRJAN | Federation of Industries of the State of Rio de Janeiro |
| IBAMA | Brazilian Institute for Environment and Renewable Natural Resources |
| IBGE | Brazilian Institute of Geography and Statistics |
| IDB | Inter-American Development Bank |
| JADECO | Japan–Brazil Agricultural Development Corporation |
| JICA | Japan International Cooperation Agency |
| JIRCAS | Japan International Research Center for Agricultural Sciences |
| MAPA | Ministry of Agriculture Livestock and Supply |
| MG | Minas Gerais (Brazilian state) |
| MT | Mato Grosso (Brazilian state) |
| NIS | national innovation system |
| PADAP | Alto Paranaíba Agricultural Development Program |
| PCRI | Integrated Rural Credit Program |
| POLOCENTRO | Cerrado Development Program |
| PRODECER | Japanese–Brazilian Cooperation Program for Cerrados Development |
| PROFIR | Cerrado Irrigation Program |
| SIS | sectoral innovation system |
| UFG | Universidade Federal de Goiás |
| UFLA | Universidade Federal de Lavras |
| UFV | Universidade Federal de Viçosa |
| UNB | Universidade de Brasília |

# Introduction: Development of Cerrado Agriculture

*Akio Hosono, Carlos Magno, and Yutaka Hongo*

Brazil was for many years a 'monoculture exporter,' exporting mainly tropical agricultural commodities, such as coffee and sugar produced in the south and on the Atlantic coast of the country. Brazil was a net importer of grain, until 1980s when the country achieved a major breakthrough to become a net exporter of grain after converting barren land into some of the most productive agricultural fields in the world. Starting in the mid-1970s, the tropical savanna of Brazil, called the Cerrado, transformed itself in just a quarter of a century into one of the world's best-known grain-growing regions, realizing modern, upland, rain-fed grain farming in a tropical region for the first time in human history. Highly skilled and entrepreneurial family farmers were the main actors in this remarkable transformation.

Today, Brazil is one of the world's major grain-producing countries, and in 2011 exported the world's largest volume of soybeans. Dr. Norman E. Borlaug, who received the Nobel Peace Prize for his work in connection with the Green Revolution, rated the development of agriculture in the Cerrado as "one of the great achievements of agricultural science in the 20th century."[1] Crops grown in the Cerrado contribute substantially to global food security.

Furthermore, an extensive agro-industry value chain has developed in the Cerrado, generating competitive value-added exports such as meat. The development of the Cerrado would not have been possible without soybeans, especially varieties adapted to tropical conditions. However, today's Cerrado is much more than soybeans, producing a variety of crops, including grain such as maize, the feijão bean, sorghum, and most recently wheat, and other crops such as cotton, coffee, vegetables, sugarcane, and high-value fruits, in conjunction with the deepening of agro-industrial value chains' network, especially of meat and dairy

1

products. What has happened in the Cerrado, especially in the last two decades, is not the development of mono-crop agriculture, but a diversified agro-industry. This development has contributed to a significant expansion in employment opportunities in the Cerrado region, and to regional development.

The development of the Cerrado is one of the crucial factors in Brazil's impressive poverty reduction. Through the generation of jobs, the promotion of socially inclusive growth, and the improvement of nutrition and food security, Brazil has achieved substantial poverty reduction. For example, during the Lula administration (2003–2010), the proportion of Brazil's population living in poverty dropped from 30 percent to 18 percent. Control of inflation, an increase in the minimum wage, and conditional cash transfers (CCT) are among the factors that directly contributed to this improvement. Additionally, over the long-term, new employment opportunities in the Cerrado and improvements in nutrition due to decreases in the real price of grain have contributed to poverty reduction. In particular, a large migration of people from Northeast Brazil, the poorest region of the country, to take up jobs in the Cerrado region has been observed.

From the early stages of the development process in this region, environmental considerations were incorporated in the interest of sustainable development, and thus agriculture in the Cerrado can indeed be regarded as a sustainable development model that has generated food production, agro-industry value chains, and employment opportunities. In the interest of environmental and ecological conservation, many innovative technologies and institutions were developed and introduced, three of which deserve special mention: first, the improvement of cultivation technologies, of which one of the most effective has been no-tillage cultivation technology; second, among institutional-cum-technological innovations, the monitoring and surveillance of illegal logging using satellite images including those with the PALSAR microwave sensor, and the 'New Forest Code' (Novo Codigo Forestal in Portuguese) and Rural Environmental Rigistry (CAR), which are among the most advanced initiatives of this kind in the world; third, the Soybean Moratorium, promoted by civil society's efforts, contributed to the reduction of rainforest destruction, when soybean varieties fit for tropical climate were utilized in the Amazon rainforest. It worked as a ban on non-eco-friendly soybean production in newly created fields in the Amazon.

The success of agriculture in the Cerrado meant that, for the first time in history, a country with a tropical climate was not obliged to depend on a tropical commodity monoculture structure, but was instead able to produce grain and achieve food security without relying upon imports. The Cerrado development could therefore provide a valuable model for contemporary developing countries struggling to attain nutrition and food security, to create value chains and jobs, to generate socially inclusive growth, and to achieve sustainable development.

The experience of creating agriculture in the Cerrado could also be valuable from the perspective of the debate surrounding economic transformation. Industrial strategy and economic transformation have been attracting renewed attention of late, and several studies in the past decade or so have focused on industrial development, especially the upgrading and diversification of industrial structures, as a basis for sustained economic growth and development.[2]

These studies have emphasized such aspects as the accumulation of knowledge and capabilities and the creation of a learning society (Cimoli et al., 2009; Stiglitz and Greenwald, 2014); the change or exploitation of factor endowments and comparative advantages (Lin, 2012); the need to compensate for the information externalities generated by pioneer firms (Rodrik, 2007); and the adoption of pragmatic policy-making for developing countries to cope with the pressures of market orientation and globalization (Ohno, 2013).[3]

The main objective of this book is to provide insights into how the above-mentioned factors interacted in the specific case of agriculture in the Cerrado to result in a remarkable economic transformation in Brazil and especially in the Cerrado region of the country.

Part I of this book first explains how agriculture in the Cerrado was made possible. It then analyzes the progress of socially inclusive development brought about by Cerrado agriculture, such as the expansion of employment opportunities and regional development. Finally, agriculture in the Cerrado region is discussed from the viewpoint of sustainable development. The Portuguese word *cerrado* refers to 'closed land,' or land that was for many years regarded as being unsuitable for agriculture. The total area of this vast region is about 204.7 million hectares, or 5.5 times the land area of Japan (Sano et al., 2007). Most of the soils under the vegetation in the Cerrado are oxisols: deep, highly weathered soils with a low capacity for holding water. Aluminum saturation is high and acidity is very high, while levels of available nutrients are very low. All these characteristics led this land to be considered unsuitable for agriculture.[4] It should be noted that the Cerrado differs greatly from the tropical rainforest in the Amazon region, and from the *caatinga* and *sertão* found in arid and semi-arid Northeast Brazil. The Cerrado area nevertheless extends widely over a zone between the latitudes 4° north and 24° south, across the tropical rainforest region of the Amazon. Its altitude also varies, between 100 and 1,200 meters, making the climate and vegetation diverse.

How did agriculture become possible in the Cerrado, which was marginal land regarded as unfit for farming? The objective of this book is to point out the major factors in this development by analyzing it as a process of establishing and developing a new industry called 'Cerrado agriculture.' The nature of Cerrado agriculture is clearer when it is considered as a newly established industry, rather than a new regional development within Brazilian agriculture. This is because Cerrado agriculture differs greatly from conventional Brazilian agriculture not only because it became possible in areas that used to be marginal and unfarmed, but also because crops generally produced in temperate regions (such as soybeans) were put into production in a tropical region for the first time anywhere in the world, with high-levels of productivity. This new type of production was enabled by the application of technological and institutional innovations.

New industries are not typically developed overnight, and many conditions must be met to make them possible. There is a need for capital, new technologies, and the development of related industries that support industrial development, as well as a labor force that includes entrepreneurs, agronomists, and other trained industrial human resources. This book will examine how these conditions were met in the case of Cerrado agriculture, as well as how diverse cooperation between Brazil and Japan contributed to the process.

Any new industry requires a private sector investment, to generate profits and to reinvest them to ensure the self-sufficient expansion of production. For this purpose, the industry's potential to yield profits as a private-sector business needed to be demonstrated. This study regards the establishment period as having occurred over approximately ten years starting at the end of the 1970s, during which agriculture in the Cerrado region was proven to be feasible as a business. An especially important outcome during this period was the success of the first phase of the Japanese–Brazilian Cooperation Program for Cerrados Development (PRODECER). This confirmed the commercial feasibility of Cerrado agriculture, and self-supporting development by the private sector in the central Cerrado region accelerated thereafter.

The years prior to this establishment period constitute the preparatory period. The most important goals to be attained at that time were the initiation of technological development, which was imperative for establishing and developing Cerrado agriculture, and the development of human resources such as researchers and agronomists. These efforts continued through the establishment period as well. The activities of the Brazilian Agricultural Research Corporation (EMBRAPA) were essential for this technological innovation. The success of the pioneer Cerrado agriculture project, called PADAP (to be discussed in Chapter 2), which was achieved through public and private sector cooperation between the state government of Minas Gerais and the Cotia Agricultural Cooperative and which employed an innovative rural credit scheme called the Integrated Rural Credit Program (Programa de Credito Rural Integrado – PCRI), was another important factor in the success of Cerrado agriculture. The former can be regarded as the starting point of the technological innovations, while the latter was the starting point of the institutional innovations, since it served as a pilot project for the structured or systematic development of Cerrado agriculture.

Because it aimed to produce crops usually grown in temperate zones, including soybeans, maize, beans, and wheat, in the tropical savannah, this agricultural development project had no overseas reference examples. No models or relevant technology could be found in advanced countries located in temperate zones. Thus, the preparatory and establishment periods were extremely important.

The years following the establishment period, when private entrepreneurs autonomously invested in the new industry, can be referred to as the early development period. Many difficulties were expected during this period, since development, even when it has begun, can come to a halt if initial hurdles are not surmounted. In other words, the key to determining whether autonomous production can be expected to expand after the demonstration of business feasibility is whether private entrepreneurs can overcome the initial difficulties on their own. This was especially true in the case of Brazil, where a debt crisis broke out in 1982, followed by hyperinflation, high interest rates, and other trials that made the steady development of Cerrado agriculture difficult. It was during this early development period that the second phase of the PRODECER project was implemented.

Chapter 1 discusses the preparatory and establishment periods, with a focus on technological innovation, Chapter 2 the establishment and early development periods with a focus on institutional innovations, and Chapter 3 the full-fledged development period. These chapters seek to point out the factors that made it possible to establish and develop Cerrado agriculture in each of these phases. Subsequently, this study provides two perspectives on Cerrado agriculture: first, in Chapter 4, an analysis of the progress of socially inclusive growth, including the diversification of production, deepening of agro-industrial value chains, expansion of employment opportunities, and regional development brought about by Cerrado agriculture; and second, in Chapter 5, an examination of Cerrado agriculture from the standpoint of sustainable development. The former attempts to demonstrate how Cerrado agriculture contributed to social development in Brazil, focusing on benefits to people in local low-income communities, while the latter aims to look at the environmental and ecological considerations that have been taken into account during the process of developing Cerrado agriculture.

These five chapters compose Part I of this book. Part II goes on to discuss the critical roles played by institutions and programs in the successful development of Cerrado agriculture. Chapter 6 explains how the Brazilian Agricultural Research Corporation (EMBRAPA) achieved outstanding technological innovations which enabled the conversion of the Cerrado from a vast, barren land into one of the world's top food baskets, focusing on the institutional building based on the 'EMBRAPA model.' Chapter 7 discusses the accomplishments of the Cerrado Agricultural Research Center (CPAC) of EMBRAPA, which played a major role in the technological innovations required for Cerrado agricultural development. Chapter 8 appraises how environment-friendly land use has been achieved, in terms of ecological and environmental conservation. Chapter 9 analyzes the role of PRODECER in the process, from the start-up to the full-fledged development of Cerrado agriculture. Finally, Chapter 10 focuses on the critical role played by CAMPO (Agricultural Promotion Company), a public–private entity, in the implementation and coordination of PRODECER in the field.

## Notes

1. httl://www.worldfoodprize.org/laureates/Past/2006.htm
2. We use the terms 'industry' and 'industrial sector' very broadly to refer not only to the manufacturing sector but also to agriculture, agro-industry, agri-business, transport, logistics, and other activities. Similarly, industrial strategy refers not only to narrowly defined 'industrial policy' targeted at manufacturing but also to other policies such as education policy, fiscal policy, financial policy, trade policy, and labor policy, which encourage the development of the aforementioned productive activities.
3. Regarding these studies, see Hosono (Forthcoming).
4. Botanically, *cerrado* is defined as "a generic term referring to vegetation where the land is continuously covered by poaceous plants, with intermittent bushes of twisted shrub with suberous and thick bark; or it refers to a generally flat area covered by this vegetation and having a long dry period" (Editora Nova Fronteira, 1987).

# References

Cimoli, Mario, Giovanni Dosi, and Joseph Stiglitz (eds) (2009) *Industrial policy and development: The political economy of capabilities accumulation* (Toronto: Oxford University Press).

Editora Nova Fronteira (1987) *Diccionario Aurelio* (Sao Paulo: Editora Nova Fronteira).

Hosono, Akio (Forthcoming ) 'Industrial strategy and economic transformation: Lessons from five outstanding cases,' Norman, Akbar and Joseph Stiglitz (eds.) (Forthcoming)*Industrial policy and economic transformation in Africa* (New York: Columbia University Press).

Lin, Justin Yifu (2012) *New structural economics: A framework for rethinking development and policy* (Washington, D.C.: World Bank).

Ohno, Kenichi (2013) *Learning to industrialize: From given growth to policy-aided value creation* (New York: Routledge).

Rodrik, Dani (2007) *One economics, many recipes: Globalization, institutions, and economic growth* (New Jersey: Princeton University Press).

Sano, Edson E., Roberto Rosa, Jorge Luís Silva Brito, and Laerte Guimarães Ferreira (2007) 'Mapeamento do cobertura vegetal do bioma Cerrado: Estrategias e resultados,' Documento 190, November 2007 (Planaltina, DF: EMBRAPA Cerrados).

Stiglitz, Joseph and Bruce Greenwald (2014) *Creating a learning society: A new approach to growth, development, and social progress* (New York: Columbia University Press).

# Part I

# Development of Cerrado Agriculture

# 1
# Technological Innovations That Made Cerrado Agriculture Possible

*Akio Hosono and Yutaka Hongo*

## Introduction

For the development of Cerrado agriculture, three technological advancements were essential. First, the soil needed to be improved and new crop varieties suited to tropical zones developed. These constituted the core technological innovations needed to develop Cerrado agriculture from practically nothing. Second, new technologies and practices needed to be effectively transferred to a large number of medium-sized farms, as family farmers were the main actors in Cerrado agriculture, rather than the limited number of companies that are often responsible for launching new industries. Third, a solid and highly effective institution needed to be involved with the continuous technological innovation required for Cerrado agriculture in order to geographically scale up, diversify, and deepen its agro-industrial value chains. To address this need, EMBRAPA (Brazilian Agricultural Research Corporation) was created and subsequently expanded, and a strong innovation system was developed. The following sections of this chapter discuss each of these crucial factors, which enabled Cerrado agriculture to be successful.

## 1.1 Scientific discovery that changed the view of the Cerrado

The Cerrado region was considered to be a marginal area for agriculture until the 1960s, and thus had been largely abandoned for a long period. Land use included only subsistence agriculture along the rivers and some beef cattle grazing on native pastures, with stocking rates ranging from 5 to 20 hectares per cow.

This conventional view of the Cerrado was called into question in the early 1940s. In 1942, two botanists from the University of São Paulo, Professors Felix Kurt Rawitscher and Mário Guimarães Ferri, began to doubt the commonly held belief that the state of the vegetation in the Cerrado was caused by dryness. They discovered that many of the native plants did not defoliate even in the driest season; on the contrary, some of them even began to bud or sprout. Plants in dry areas usually have only a small number of pores on the underside of their leaves in order to avoid evaporation, but this was not true of the native plants of the Cerrado. Detailed studies were then initiated to investigate the causes of the Cerrado's peculiar vegetation. Studies that began in botany rapidly expanded to include geology, physiography, soil science, meteorology, zoology, and agronomy, with regional universities and research institutes participating alongside the University of São Paulo. By 1952, the theory that attributed the Cerrado's vegetation to the soil instead of the lack of rainfall came to be gradually supported (Alvim and Araujo, 1952).

In 1959, Ferri and others proposed the concept of oligotrophic sclero-morphism, a theory that attributed the state of vegetation in the Cerrado to the chemical factors of the soil rather than lack of rainfall (see Ferri, 1971). This theory can be summarized as follows: Cerrado soil is among the oldest soils in the world, having existed for at least a million years, but nutrients had been leached from it during that period, bringing it to an exhausted state. This severely worn-down soil is classified as latosol (oxisol and ultisol in the US taxonomy) and is typical of Cerrado soil, occupying 46 percent of the area. It also has high acidity and aluminum toxicity. Plants capable of withstanding these severe conditions are not capable of fully synthesizing protein from carbohydrates, which have accumulated in great quantities within the plants due to their active photosynthesis in the sub-tropical region. As a result, these plants – the peculiar trees of the Cerrado – have thick leaves and bark.

This scientific discovery completely changed the view of the Cerrado as marginal land. From that point on, research began on how to make use of the Cerrado for agriculture, which led to the discovery of its agricultural advantages, including: (1) the rainy season, during which annual rainfall ranges from 500 mm in the Northeast to 2,300 mm in the sub-Amazonian areas; (2) the deep and friable soil; (3) the potential to improve soil conditions using limestone and fertilizers; (4) the vast, flat land, which is ideally suited to large-scale mechanized farming; and (5) the shrub vegetation, which is easy to clear so that land can be reclaimed, thereby reducing initial cultivation costs (agricultural land development), especially compared with the tropical rainforest in the

Amazon. As the Cerrado's natural vegetation was considered to be much poorer than the Amazon's, Cerrado agriculture drew little opposition from environmental standpoints, and this was one of the factors that accelerated its development.

## 1.2 Creation of the Brazilian Agricultural Research Corporation (EMBRAPA) to foster technological innovation

Thus, the potential to transform the Cerrado from 'sterile' land into arable land was gradually revealed. Yet, it was not easy for private businesses and farmers to independently achieve the technological innovations necessary for this purpose. Some pioneering farmers did have positive outcomes and attained success, and this contributed a great deal to the overall efforts, as will be discussed in Chapter 2, but the Brazilian government judged that it was necessary to establish an organization dedicated to fostering the necessary technological innovations. The Brazilian Agricultural Research Corporation (EMBRAPA) was established in 1973 and did in fact achieve much success: recent discussions on the Cerrado point out that EMBRAPA's greatest contributions were soil improvement (the control of soil acidity and the improvement of soil fertility) and the development of new varieties of crops such as soybeans, maize, rice, beans, and wheat that were better suited to tropical zones (FAO and The World Bank. 2009, The Economist, 2010; Inter-American Development Bank (IDB), 2010; Correa and Schmitt, 2014).

With the creation of EMBRAPA, Brazil addressed the high cost and risk of research and development for agriculture in general and for Cerrado agriculture in particular. Investment in learning, knowledge, and technology is highly risky, and risk markets are absent (especially in developing countries), which in turn discourages such investments (Norman and Stiglitz, 2012). Further discouraging private investment is the general feeling that the technology and knowledge required for agriculture should be a public good. Though such technological innovations require high investment, investors cannot appropriate the knowledge and technology, which should instead be disseminated freely to farmers for agricultural development. EMBRAPA, a public entity established in 1973, effectively invested in knowledge and technology and provided them as a public good to farmers, who started to cultivate land in the Cerrado.

Analyzing how technological innovations were made possible in the Cerrado can provide clues to the factors behind EMBRAPA's successful research and development. Established in 1975, EMBRAPA's Cerrado Agricultural Research Center (CPAC, also called "EMBRAPA Cerrado") achieved success very early as an innovative research institution. The CPAC realized technological innovations for soil management and the improvement of crop breeding, both of which were fundamental to the development of Cerrado agriculture. Although the CPAC is a research institute affiliated with EMBRAPA, it is no exaggeration to say that this institute virtually *was* EMBRAPA at that time, as EMBRAPA had been established to support agricultural development mainly in the Cerrado. The Japan–Brazil Agricultural Research Cooperation Program in the Cerrado was started at the CPAC by EMBRAPA and the Japan International Cooperation Agency (JICA) in 1977.[1] This technical cooperation program began prior to the Japanese–Brazilian Cooperation Program for Cerrados Development (PRODECER), which will be discussed in Chapter 2, and served as technological preparation for PRODECER.[2]

The research begun by EMBRAPA in 1973 progressed steadily, making it one of the largest agricultural research institutes in the southern hemisphere, and one of the largest tropical agricultural research institutes in the world. At the end of 2011, there were 9,803 people working for the corporation, 2,389 of whom were researchers, with 1,959 of these possessing PhDs. In comparison, only three researchers with doctorates had been with EMBRAPA at its founding in 1973. Since then, EMBRAPA has dispatched 3,000 people to developed countries to work or study, and it now has 47 affiliated research centers and is highly appreciated overseas for its distinguished research.

## 1.3   Soil improvement: a core technological innovation

Most of the soils under Cerrado vegetation are oxisols. They are deep, highly weathered soils with a low capacity for holding water. The soils' acidity was very high and their aluminum saturation was high, while levels of available nutrients were extremely low. These characteristics led to the conclusion that these soils were unsuitable for agriculture.[3] Therefore, the first research conducted by the CPAC was on soil improvement, using cutting-edge soil analyzers, which helped to reduce the soil analysis period from several days to several hours. One of the greatest obstacles to the development of Cerrado agriculture was the soil itself, so these analyzers had a great impact on the project. They accelerated soil research and increased accuracy, and as a result the number of papers related to soil science published in academic journals by the CPAC increased rapidly.

At nationwide Cerrado symposiums regularly held in Brazil, the CPAC presented its most up-to-date soil research results, which made a great contribution to the development of Cerrado agriculture. These results were compiled in literature such as the book *Soils of the Cerrado: Technologies and strategies of management*, published by the CPAC in 1989. Many researchers were involved in the soil research, with especially noteworthy contributions by Dr. Wenceslau J. Goedert and Edson Lobato. Goedert was appointed as the first Director of the CPAC and established the basic direction of its soil research. He analyzed Cerrado soils, contributed to the establishment of the methodology to convert them into agricultural fields, and was the editor of the above-cited book. Lobato received the 2006 World Food Prize for his outstanding role in helping to transform the Cerrado into productive cropland. He worked to enhance soil quality and counteract water stress, and collaborated with farmers and extension technicians to implement the technologies and practices related to soil fertility and soil management pioneered at the CPAC. A book that he edited in 2004, *Cerrado: Soil Correction and Fertilization* became a standard reference for both farmers and researchers.

Brazilian and Japanese researchers were convinced that detailed data needed to be collected to clarify the diverse characteristics of the various areas in the Cerrado. Carlos Magno Campos da Rocha, who served as the Director of the CPAC and later as the President of EMBRAPA, noted that detailed studies on the soil of the Cerrado began with the start of Japan's cooperation.[4] In fact, the initiation of research cooperation between Brazil and Japan triggered the shift to the establishment period.

Until around 1978, during the preparatory stage of Cerrado agriculture, there were only about 700 research papers on the Cerrado, most of which dealt with botanical issues such as features of the vegetation. This situation completely changed when the establishment period for Cerrado agriculture began, partly due to the creation and strengthening of the CPAC. At this time there was a surge in the number of agronomical research papers, which placed greater emphasis on soil and crop production. This was the crucial difference between the preparatory and establishment periods of Cerrado agriculture. The number of research papers on the Cerrado increased dramatically during the first five years of Brazil–Japan research cooperation, and the ratio of papers on agricultural technologies rose to 95 percent. Most of the CPAC's Brazilian researchers later studied at the Tropical Agricultural Research Center in Tsukuba (later reorganized as JIRCAS), Japan, further underscoring the importance of this cooperation.

## 1.4   Soybeans fit for the tropical climate: a technological breakthrough for Cerrado agriculture

Besides soil improvement, another breakthrough was the success in developing new crop varieties that were adapted to the tropical environment, with a focus at first mainly on soybeans. Prior to this, it was considered impossible to develop large-scale agriculture in the Cerrado because there were no crops well suited to the environment there. Even acid-resistant rice, which was considered an exception to other crops, which cannot resist the high acidity of the Cerrado soil, could not be pollinated during the 'boot stage,' or ear formation stage, of development due to the short dry season, known locally as 'veranico,' which occurs during the rainy season in the Cerrado and which causes a high-risk of empty husks. Moreover, as soil deterioration was caused by continuous cropping upland, rice production was not viable and had to be abandoned in two to three years.

Under these circumstances, much was expected from soybeans. The soybean is a legume which can contribute to the fertilization of soil through the fixation of atmospheric nitrogen with root nodule bacteria. It also gradually accumulates organic matter in the soil. It was the soybean that started the process to help the land to become fit for agriculture, accordingly. As a result, soybeans were important for the agricultural development in the Cerrado, not only as the first economically viable crop, but also because they were indispensable in making the soil in the Cerrado fit for agriculture. In fact, the soybean was a doubly pioneering crop for the Cerrado, since the development of the Cerrado required an initial substantial investment and therefore even the pioneering crop needed to be economically profitable during the soil remediation stage. A crop which was stable and reached a reasonable level of productivity during its early years was absolutely necessary, and this was the soybean.

There were two big obstacles to growing soybeans in the Cerrado. The first was its photoperiodism. Soybeans bloom and sprout by sensing differences in day lengths (photoperiod). Soybeans were originally grown in temperate zones, where plants are generally exposed to 14 to 16 hours of sunlight per day around the summer solstice and then fewer and fewer hours. This shift to shorter days prompts flowering and fruiting.

This makes soybean cultivation difficult in tropical regions, and even more so in the lower-latitude areas in the Cerrado, where the length of the day is nearly constant year-round. If soybeans grown in the temperate zone in the south of Brazil were introduced in the Cerrado in the tropical region at low latitudes – even if the seeds were sown in mid-October when the rainy season starts – the days would not be long enough to promote growing, which results in flower bud differentiation and dwarfing. Alternatively, the plants could collapse. In these circumstances, the use of combines (harvesters) would not be possible.

The second obstacle was the soil. The soil in the Cerrado was highly acidic, with a pH level of 4 to 5, and nutrients were extremely deficient. Even if farmland were reclaimed by cutting down shrubs, and lime and phosphorus fertilizer provided, soil would not be improved to the point that soybeans would start to grow immediately. Even the application of three to four times the usual amount of bradyrhizobial inoculants would produce hardly any effect. It would take at least three years (three crops) for the soil in the Cerrado to become fit for agriculture.

Among researchers who specialized in breeding, Dr. Plínio Itamar Mello de Souza deserves special mention[5] as the first to successfully develop the revolutionary varieties of soybeans suited to the tropical region. De Souza devoted his work at the CPAC to the breeding of a tropical soybean variety which could be cultivated immediately after the completion of the farmland reclamation in the Cerrado. To succeed in this development, he collected 3,000 soybean varieties from the southern United States, the Philippines, Japan, and other parts of the world, selected those with a low sensitivity to changes in day length, then selected those that grew tall in tropical regions, and finally crossbred them with varieties with high yields. In cooperation with the EMBRAPA Soybean National Research Center, he selected 'Lo75–2760' as a potential variety and further improved upon it. In 1980, Dr. Souza finally completed the first soybean variety for cultivation in the Cerrado. EMBRAPA named this variety 'Doko.'[6]

Doko became very popular from the beginning of the development of the Cerrado for several reasons, such as its (1) adaptability to the reclaimed farmland soil of the Cerrado, (2) stable productivity and economic performance, (3) height, which was sufficient for the introduction of harvesters, (4) adaptability to environmental changes, and (5) low dependence on rhizobium, among other factors. Every year, 200 to 250 new soybean varieties are developed in Brazil. As the land becomes more arable, new varieties replace the old ones, and most varieties last five to seven years. Doko, on the other hand, has survived 20 years since its development. At one point the variety boasted the world's largest cultivation area.

## 1.5　Technological innovations that enabled Cerrado agriculture to scale up, diversify, and develop its agro-industrial value chains

In terms of the geographical scaling-up of Cerrado agriculture, it is noteworthy that many of EMBRAPA's research centers were established in Brazil's Central-West region, in view of the fact that the development of Cerrado agriculture was at the core of its mission (and it continues to address issues related to the Cerrado to this day). Brasilia, the capital of Brazil, is located in this region and is at the heart of the Cerrado. One important feature of EMBRAPA's organization was that it had three types of research center: regional research centers, crop research centers (including livestock and fishery products), and research centers for working on cross-cutting subjects.

Among these, the Cerrado Agricultural Research Center (CPAC, also called "EMBRAPA Cerrado") is located 30 miles northeast of Brasilia. EMBRAPA also has headquarters in Brasilia, where EMBRAPA Vegetables, EMBRAPA Coffee, EMBRAPA Agro-energy, EMBRAPA Studies and Training, and EMBRAPA Genetic Resources and Biotechnology are also located. EMBRAPA Western Region Agriculture, EMBRAPA Rice and Beans, and EMBRAPA Beef Cattle are also located in the Central-West region of the country. EMBRAPA Maize and Sorghum is located in the Southeast region, but is, from the standpoint of ecological systems, within the Cerrado area. EMBRAPA Soybeans, EMBRAPA Wheat, and EMBRAPA Dairy Cattle are located in the South and Southeast regions; however, these research centers also study Cerrado agriculture. This illustrates how EMBRAPA has conducted research on a large-scale, from multiple angles, and in a decentralized manner; it also shows that the development of Cerrado agriculture is at the core of the organization's focus.

EMBRAPA Centers specialized in specific agricultural crops and livestock production, contributed significantly to the diversification of Cerrado agriculture. The following are some of outstanding examples. EMBRAPA Beef Cattle succeeded in breeding a beef cattle variety fit for the Cerrado, developed technology for raising and fattening the cattle, introduced from African countries superior varieties of pasture such as Brachiaria and Panicum (Capim Colonião), and improved the practice of pasture management. EMBRAPA Coffee developed varieties of coffee fit for the Cerrado and improved the technology for its cultivation, as coffee was not previously a crop of this biome. The organization also developed technology to get rid of harmful insects and to irrigate coffee plantations.

The achievements of EMBRAPA Vegetables were equally remarkable. It developed varieties of vegetables fit for the Cerrado through breeding and the improvement of technologies for cultivation. It also developed varieties of tomato and other agricultural inputs for processing. EMBRAPA Vegetables' success in developing new varieties, such as a variety of carrot called Cenoura CV.Brasília and a pumpkin called Abóbora Tetsukabuto is well known.

In addition to EMBRAPA, many research institutes, universities, and other institutions participated in research and development for Cerrado agriculture. These included Empresa de Pesquisa Agropecuária de Minas Gerais (EPAMIG), Empresa Baiana de Desenvolvimento Agricola (EBDA), Universidade Federal de Viçosa (UFV), Universidade Federal de Lavras (UFLA), Universidade de Brasília (UNB), Universidade Federal de Goiás (UFG), and Fundação de Mato Grosso.

As its name indicates, EMBRAPA is a national research institute for agricultural technologies. It needs to secure a sufficient portion of its budget from the central government in order to fulfill its mission. Therefore, the importance of its research needs to be fully recognized by the government and parliament. In other words, its budget will not be allocated if EMBRAPA fails to continue to demonstrate that it has achieved useful research results. The success of Cerrado agriculture has therefore been vital to the future of this organization.

EMBRAPA set the development of Cerrado agriculture as its core mission and achieved success, thereby establishing its position, and thereafter steadily secured its research budget while maintaining political neutrality; by securing its research budget, EMBRAPA obtained further research results, which reinforced its position. The view of Dr. Elieseu Alves, a former President of EMBRAPA, could be thusly summarized:What solidified the position of EMBRAPA was the achievement of transforming the Cerrado into a modern agricultural region. EMBRAPA's contributions are at the core of Cerrado agriculture, and society recognized that its involvement is vitally important for its success. (see Alves 2012a for more details). This process has been crucial for the continuous development of technologies necessary for the scaling-up of Cerrado agriculture, its diversification, and the deepening of its agro-industrial value chains.

## 1.6   Factors behind the EMBRAPA's success[7]

Analyzing the factors behind EMBRAPA's success reveals some clues as to how to develop organizations capable of R&D activities suited to a country's demands, which at the same time can generate technological innovations, develop human resources, and produce outstanding results similar to those in the Cerrado.

Dr. Eliseu Alves, who has analyzed the factors behind the success of the EMBRAPA model,[8] has pointed out that, above all, it aimed at social interaction. EMBRAPA placed particular emphasis on building close relationships between researchers and agricultural producers. Dr. Alves is a specialist in technological dissemination, and he believed that research and technological development based on the needs of producers would lead to increased productivity. He notes that close relationships between researchers and producers were especially important for disseminating and effectively utilizing the results of research and for identifying study issues that should be adopted as research priorities.

Each researcher was urged to interact frequently with the public and to behave like an entrepreneur disseminating technology. Producers knew where to go when they needed information, while EMBRAPA researchers accurately recognized their duties. The producers who first engaged in Cerrado agriculture were migrants from the country's South and Southeast regions. They sold their farmland in their hometowns, purchased larger land lots in the Cerrado, and became engaged in Cerrado agriculture. The important factor was that these migrants were experienced in agricultural administration,[9] so they immediately began applying technologies developed by research institutes like EMBRAPA.

Fostering researchers has been an especially important objective sought by EMBRAPA since its foundation. It was believed that human resources were essential for EMBRAPA to succeed. Career systems for researchers were established, and salaries and various other conditions were set so that researchers would take pride in their profession. A reward system was established through which individual researchers or groups of researchers were promoted on the basis of their achievements. Training programs were made available to them at top research centers in more advanced countries. This was an effective way of training researchers and at the same time served as a powerful incentive for them. Many researchers went abroad to study, and obtained graduate degrees. The high regard in which EMBRAPA is held as a research institute made it possible to recruit excellent young researchers.

Each EMBRAPA research unit was assured of having the necessary critical mass of researchers, and would be given a working environment where researchers could be trained and contribute to creative work and to society. Researchers were urged to practice teamwork and to share their knowledge with their teammates.

In order to realize the above goals as a research organization, especially with regard to the treatment of and career systems for researchers, EMBRAPA made an important decision when it was founded not to apply the rules usually followed by administrative agencies. This enabled it to flexibly carry out budget and personnel management, plan strategies, assess and disseminate the results of research, execute the budget, and assure transparency. Especially with regard to the career development of researchers and the assessment of their research results, the organization adhered to the conventions on personnel management of private-sector businesses instead of the regulations for government officers (Chapter 6).

The founders of EMBRAPA believed that consistent and steady efforts were needed for research activities, and that independence from politics needed to be maintained for this purpose. To this end, EMBRAPA implemented measures to ensure transparency in its operations, in addition to the legal measures already noted, so as to avoid excessive political intervention by the government and the imposition of various restrictions. Information disclosure was promoted and EMBRAPA held public selections of the people who would assume the office of Director at each of its research centers. Other governmental institutions regarded EMBRAPA's highly independent systems as a reference model.

Another characteristic of EMBRAPA was its effective use of the media for publicity activities. This included promptly disseminating its research results to agricultural producers, and continually winning the support of the government, agri-business, and the general public, all of which were essential for securing the budget EMBRAPA required. From its early years, EMBRAPA had a clear vision and determined the roles it should perform, and its researchers shared these views.

Tremendous efforts were made during the early years to communicate the "EMBRAPA spirit" to researchers at meetings and seminars, and on other occasions. Dr. Alves notes that today, EMBRAPA is well known and it is therefore easy to develop the EMBRAPA spirit, but its importance should never be underestimated (Chapter 6). Part of this spirit is formed by the emphasis on developing researchers and on the close interactions with society discussed above. The high priority that EMBRAPA gives to Cerrado agriculture is also indicative of how it recognizes its mission.

The EMBRAPA spirit is closely related to the institution's scale and organizational structure, as well as the locations and the roles of its affiliated research centers. Its scale must be in proportion to Brazil's extensive area of land: otherwise, it will not be able to respond to the country's needs. Thus, it did not attempt to develop a small-scale organization like those in universities and research institutes elsewhere; instead, it established a large-scale, diverse, and decentralized organization, driven by problem-solving.

Although further in-depth analysis is needed, the above discussion suggests that EMBRAPA and the EMBRAPA model could be considered a model of an effective public research organization from the viewpoints of management and organization theories.[10] In the EMBRAPA model, most of the conditions that are considered essential for an organization to be effective are met: clear strategy (mission, objectives, and goals), staff incentives, group cohesiveness, organization structure, internal communication, leadership, and culture. Autonomy of management, insulation from political influences, transparency, and good relationships both with stakeholders and the general public were also crucial for EMBRAPA's effectiveness in the political economy of Brazil.

The public goods (and services) EMBRAPA produced and delivered to the beneficiaries – knowledge and technology – are different from normal goods and services. This unique aspect of the organization is closely related to the issue of technological innovation and dissemination, or a national system of innovation, to be discussed in the following section.

## 1.7 Formation of a national or sectoral system of innovation and dissemination

Innovation is now widely recognized as essential for growth. Therefore, the creation of a national system of innovation is considered crucial, because the creation, diffusion, and application of knowledge that takes place through interactions among various actors of an innovation system is the key to growth (Gu et al., 2012, p. 195). While a national innovation system (NIS) and sectoral innovation systems (SIS) share a perspective that economic change is evolutionary, and that multiple actors are involved, NIS is more aggregative and is particularly oriented to broad national characteristics (Malerba and Nelson, 2012, p. 5).

From the perspective of the innovation system, EMBRAPA could be considered one of the most important hubs or coordinators of the agricultural SIS and a part of NIS. EMBRAPA established and led the National System of Agricultural Research (Ministry of Agriculture, Livestock and Supply of Brazil and JICA, 2002, Part 4, p. 26). This system has "an excellent structure and aims at technological development and its diffusion in an efficient and effective manner" (Ibid.). The system was designed for collaboration and exchange of technical information among more than 400 organizations over the whole country, including state research companies, university research organizations, and agricultural research departments in private companies.

Figure 1.1 is an adapted version of the National System of Research and Dissemination framework as identified in the Joint Study on Japan–Brazil Cooperation Programs for Cerrado Development (Ministry of Agriculture,Livestock and Supply (MAPA)(Brazil) and JICA, 2002). At national level, EMBRAPA and the Secretariat of Rural Support and Cooperatives Support of the Department of Infrastructure and Rural Extension of the Ministry of Agriculture, Livestock and Supply (MAPA) are at the center of the system coordinating both national- and state-level research and dissemination organizations (see the first level of Figure 1.1). As for Brazilian Enterprise for Technical Assistance and Rural Extension (EMBRATER), its mission and dissolution are explained below.

At state-level, departments of agronomy of private and public universities, state companies of agricultural research, and state experimental fields are the main institutions for research and dissemination (see the middle level of Figure 1.1). At the level of producers (farmers and private companies), CAMPO (a bi-national public–private entity created for the implementation and coordination of PRODECER); cooperatives; grain, seed, and fertilizer companies; consulting companies; and agricultural research foundations in states and municipalities participate in the system (see lower level of Figure 1.1). Here, the CPAC is highlighted for its research and dissemination activities, which bridged national-, state-, and producer-level stakeholders.

Understanding how this system was set up, it is important to examine how it actually worked. In particular, how were the groundbreaking technologies cited in Sections 3 and 4 spread? How did the pioneers of Cerrado agriculture improve their technological capabilities, after settling in an area they once believed to be sterile? This section will explore these questions.

As mentioned above, many of the farmers who migrated to the Cerrado from southern Brazil had experience in agricultural production and were positive about adopting new technologies. To support this, EMBRAPA placed particular emphasis on building close relationships between researchers and agricultural producers, as discussed above, for disseminating and effectively utilizing the results of research, and for identifying research priorities. This was an extremely important part of EMBRAPA's vision from its inception, since it was believed that research and technological development based on the needs of producers would lead to increased productivity.

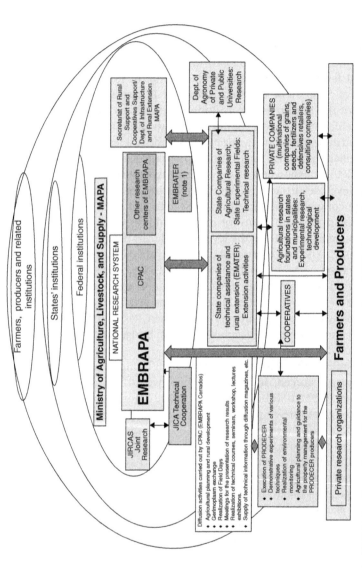

*Figure 1.1* National system of agricultural technology research and diffusion and relationships among the country's federal, states' and producers' institutions

*Note:* In 1992, EMBRATER was closed and its mission was transferred to EMBRAPA and, afterwards to MAPA.*Source:* Authors. Based on Ministry of Agriculture, Livestock and Supply (MAPA) (Brazil) and Japan International Cooperation Agency (JICA), 2002. *Japan–Brazil Agricultural Development Cooperation Programs in the Cerrado Region of Brazil: Joint Evaluation Study General Report.* Brazilia and Tokyo: MAPA and JICA.

The CPAC played an important role in achieving this vision. It promoted the development of technology and carried out experimental research and surveys about challenges that could arise during the agricultural development process in the Cerrado region. The CPAC also carried out joint research with other EMBRAPA research centers and with state experimental fields with which it maintained cooperative relationships. This activity was necessary to validate the agricultural technology developed by the CPAC itself and/or adapt it to the various socio-economic, climatic, edaphic, and topographic conditions (Ministry of Agriculture,Livestock and Supply (MAPA) (Brazil) and JICA, 2002, Part 4, p. 15). Consequently, the state experimental fields provided an important bridge between the CPAC and farmers. Within PRODECER areas, experimental fields were normally implemented by the CPAC on land provided by farmers.

It is important to understand that this process enabled stakeholders to adapt breakthrough technological innovations to local conditions, tailoring them to create locally appropriate technologies and practices. This aspect is highlighted by the Joint Evaluation Study on Japan–Brazil Agricultural Development Cooperation Programs in the Cerrado Region of Brazil (Ministry of Agriculture, Livestock and Supply (MAPA) (Brazil) and JICA, 2002), which could be summarized as follows: At federal level, basic technologies were developed. At state and farmer levels, the technologies were adapted to each producing region on an experimental basis. Farmers made efforts to improve their technical capabilities and to increase the exchange of information through visits to farms and experimental fields. There were also research foundations sponsored by states and municipalities, in which private companies participated. During site visits, the research and development activities that were being carried out intensively by some of these foundations could be observed. These foundations were developing, through joint research with the cooperatives, cultivation techniques and agriculture management (Ibid., p. 17).

The final outcome of this process, the adaptation of agricultural techniques to different regions – whereas at the initial stage of Cerrado agricultural development, there had been almost no production technologies appropriate to the Cerrado – allowed the increase of agricultural productivity and of production volume. Furthermore, the technologies developed through this process were shared not only with institutions of research and diffusion, but also with cooperatives and private enterprises that specialized in seeds, fertilizer, and insecticides, which in turn facilitated the diffusion of such technologies all over the Cerrado.

Other public and private institutions played important roles in the dissemination of innovative technologies and practices developed for Cerrado agriculture. For example, COTIA, the largest agricultural cooperative in Brazil, contributed greatly to the process of technological dissemination. The Brazilian Enterprise for Technical Assistance and Rural Extension (EMBRATER) was initially in charge of disseminating technologies developed by EMBRAPA and other organizations (see upper section of Figure 1.1), but it was dissolved as a part of administrative reform and deregulation policies in 1992.

In PRODECER, the growth pole strategy was adopted at Cerrado frontiers, as will be discussed in Chapter 2. COTIA and other cooperatives provided detailed technological support for individual farmers, contributing greatly to raising their technological capabilities. The members of cooperatives were enthusiastic and motivated. CAMPO,[11] a public–private entity established jointly by Brazil and Japan for the implementation and coordination of PRODECER, had staff stationed at each project site and helped to solve technological problems in daily activities as well as collecting information so that CAMPO headquarters could promptly and accurately assess ongoing issues.

CAMPO fulfilled its expected function by greatly contributing to technological dissemination through technical assistance and rural extension to farmers. It stimulated the implementation of experimental fields within each project area and carried out joint experiments together with the CPAC and other research organizations, aiming at validating the technologies adopted within each project area. For example, experiments with new technology were carried out at the CAMPO experimental field in Paracatu, through cooperation among CPAC, EPAMIG (Agricultural Research Corporation of Minas Gerais State) and Japanese experts.

In addition, the integrated credit scheme introduced by PRODECER to provide financial support together with technical guidance, which it had inherited from PADAP,[12] was highly effective in aiding the dissemination of advanced technologies among farmers. For this purpose, CAMPO prepared "Manuals for Technical Guidance to Producers" together with EMBRAPA and state research and extension organizations, among others. These manuals prepared for different PRODECER sites were used to help farmers design individual plans, and to define parameters for the technical departments of financial institutions for the analysis and approval of agricultural credit to farmers.

This integrated credit scheme had two innovative features which were not found in conventional credits to agriculture in Brazil. First, it offered finance for a comprehensive production package for farmers, instead of credits, for example, for the purchase of tractors or other specific purposes, in which the tractors or other physical assets were used as collateral. The integrated credit introduced in PRODECER was a package which included financing to buy land, improve soil, buy machinery, construct houses and storages, and cover the production costs and expenditures necessary to start life in the new settlement. Second, the integrated credit established as its indispensable conditionality the introduction of technologies from the Manual for Technical Guidance to Producers which compiled the most advanced technologies adapted to the local conditions of the place where farmers settled. Through this approach, the use of the cutting-edge technologies and practices developed specifically for the Cerrado was guaranteed in PRODECER settlements, which became demonstration fields for new entrants to the areas surrounding PRODECER sites. This process will be further elaborated in Chapters 2 and 10.

In addition to the support and coordination provided by CAMPO, the most important factors for the spread of technological innovations included the organizational power of agricultural cooperatives, and the high educational standards and knowledge of agriculture of the young, second-generation Japanese–Brazilians and other farmers from southern Brazil who migrated to the Cerrado in the early years, and above all their passion and hard work in response to new challenges.[13] Thus, the necessary human resources were assembled for the totally new industry of Cerrado agriculture, and their technological capabilities were improved in order to ensure their success.

## Concluding remarks

The Cerrado, a vast area in the center of Brazil, was traditionally regarded as being unfit for agriculture. Cerrado agriculture therefore had to be developed from scratch. This chapter analyzed how the technological innovations crucial to this development were achieved. In doing so, it aimed to identify the factors behind this success, including the creation of an effective institution devoted to the continuous development of technological innovations and their dissemination, as required by the changing needs of Cerrado agriculture.

The genesis of Cerrado agriculture goes back to the revolutionary move away from the conventional view that the state of vegetation in the Cerrado was caused by dryness. A scientific discovery brought the understanding that this was actually caused by characteristics of the soil.

Therefore, technologies to improve the soil were considered crucial and the establishment of EMBRAPA and its Cerrado Agricultural Research Center (CPAC) accelerated research on soil. Soil improvement was a core technological innovation and a necessary condition for the beginning of Cerrado agriculture. However, the development of new varieties of crops fit for the tropical climate, starting with soybeans, was indispensable. The production of new varieties adequate for the Cerrado was a technological breakthrough for Cerrado agriculture.

The technological innovations summarized above were essential for the establishment of Cerrado agriculture. But EMBRAPA and other agricultural research institutions as well as universities continued their efforts in research and development to scale up and diversify Cerrado agriculture, and deepen its agro-industrial value chains. This chapter aimed to analyze the factors that made this process successful and highlighted the unique characteristics of EMBRAPA, as well as the roles played by cooperatives and other public and private institutions. These findings suggest important insights into the approach to investments in knowledge and technology that is crucial for the development of a new industry.

---

Box 1.1   International cooperation

International cooperation was a key policy of EMBRAPA from its beginning. In addition to JICA and JIRCAS (from Japan), which were involved from its earliest years, institutes from the United States and France, the World Bank, the Inter-American Development Bank, and the Consultative Group on International Agricultural Research (CGIAR) have worked with EMBRAPA. Dr. Alves rates Japan's cooperation highly because it focused on institution building for EMBRAPA. It was important in this regard for researchers from both Japan and Brazil to work jointly on common research issues. Japanese and Brazilian organizations were heavily involved in the cooperation (see Chapter 6 of this book). Based on the example of the cooperation with JICA, Dr. Alves identifies three conditions for successful international cooperation: (1) the cooperating organization is prepared for the challenges that the partner organization faces, understands these challenges, and is ready to move toward a solution in a collaborative way; (2) the autonomous development of the partner institution after the completion of the cooperative program is guaranteed; and (3) the collaborative efforts by the cooperating and partner institutions generate benefits for both. Dr. Alves also mentions two types of institution building: (a) boosting research capabilities without changing the existing organizational and operational structures; and (b) changing or rearranging the organizational model itself.

He also notes that, even though Brazil established EMBRAPA without relying on cooperation with other countries, JICA contributed to institution building with respect to the CPAC concerning item (b), and also contributed greatly with respect to (a) by offering machinery and laboratory equipment. He recalls that, at the time, EMBRAPA was in need of overseas support, and Western countries and international institutes offered assistance. Yet, EMBRAPA recognized the need for a new form of cooperation (comprehensive institution building), rather than the conventional research cooperation, for the new agriculture to be practiced in the Cerrado, and selected JICA as the best choice for this purpose (Chapter 6)

Dr. Alves goes on to explain that a great deal was expected of the cooperation with Japan, because Japan set the development of Cerrado agriculture as the final objective. JICA offered a package of programs in a project-type technological cooperation[14] that involved dispatching experts, receiving trainees, and providing machinery and laboratory equipment, as well as joint research. Dr. Alves worked hard to advance negotiations with Japan in the large-scale technological cooperation project, which resulted in the first phase of the Japan–Brazil Agricultural Research Cooperation Program in the Cerrado (1977–1985); moreover, he paved the way for the second phase of the program, initially as EMBRAPA's Director in charge of technologies and then as its President.

A feature article in the August 26, 2010 edition of *The Economist* reported that "[i]mproving the soil and the new tropical soybeans were both needed for farming the Cerrado" (The Economist, 2010, pp. 58–60). The Japan–Brazil Agricultural Cooperation Program addressed these two issues from the beginning, and made notable accomplishments in the early years. Former CPAC General Director, Dr. Wenceslau J. Goedert, believes that technological developments in diverse but related areas, including crop breeding, bradyrhizobium bacteria selection, crop protection schemes, cultivation, agricultural machinery adaptation and development, and agro-zoning using satellites, surged into the mainstream of soil and breed improvements to form the comprehensive technology necessary for Cerrado agriculture.[15]

In 2006, the World Food Prize was awarded to Brazil's former Minister of Agriculture Alysson Paulinelli, who had promoted Cerrado agriculture from its beginnings, and Edson Lobato, who had contributed to soil fertility research in the Cerrado.[16]

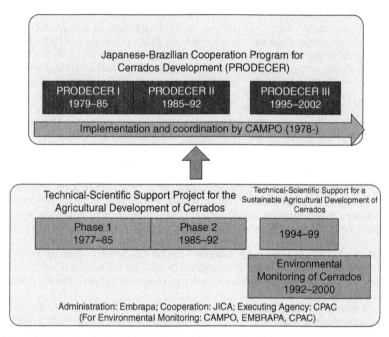

*Figure 1.2*    Technological innovation and international cooperation

*Notes:* Japan–Brazil cooperation for Cerrado agriculture development consisted of financial and technical cooperation programs. PRODECER was the major financial cooperation program, in which CAMPO played the role of the implementing and coordinating organization, and was supported by JICA/CPAC (EMBRAPA) in relation to technical cooperation. The Cerrado Irrigation Program (PROFIR) and Goias State Electrification Program were also Japan–Brazil financial cooperation programs. Research programs were also carried out jointly by EMBRAPA and the Japan International Research Center for Agricultural Sciences (JIRCAS). These included research on cultivated land in Brazil (1972–1996); a survey and analysis of the agricultural characteristics and of the direction of the technological improvement in South/Central America (1993–); "Development of Sustainable Agro-pastoral System in Sub-Tropical Zones" (1996–2002); and the "Comprehensive Soybean Research Project in South America" (1997–2006). All of the above-mentioned cooperation programs and projects came under the umbrella title "Japan–Brazil agricultural development cooperation programs in the Cerrado region of Brazil" and an evaluation study of these programs was published in 2002 (Ministry of Agriculture, Livestock, and Supply (Brazil) and JICA (2002)).

*Source:* Authors.

## Notes

1. The Japan–Brazil Agricultural Research Cooperation Program in the Cerrado refers to a series of research cooperation projects implemented by EMBRAPA and JICA, as shown in Figure 1.2. Before this comprehensive research cooperation program, the Tropical Agricultural Research Center (later renamed the Japan International Research Center for Agricultural Sciences, JIRCAS) located

in Tsukuba, Japan, began working with Brazilian researchers in 1972 on joint research on upland farming in Brazil, a collaboration that continued until 1996. JIRCAS also played an important role in JICA's technological cooperation with the CPAC, which began five years later, as mentioned above.

2. In the first phase of cooperation between EMBRAPA and JICA, JICA dispatched 50 experts and invited 33 Brazilian researchers to Japan, and machinery and laboratory materials worth 760 million yen (US$8.36 million) were made available. The areas of cooperation were diverse and included plant protection to insect control, crop development and cultivation, soil fertilization and management, moisture retention systems, agricultural meteorology, agricultural administration, and economic analysis. The Brazil–Japan Research Cooperation Program for Cerrado agriculture continued for over 20 years (Figure 1.2).

3. Botanically, *Cerrado* is defined as "a generic term referring to a vegetation where the land is continuously covered by poaceous plants, with intermittent bushes of twisted shrub with suberous and thick bark; or it refers to a generally flat area covered by this vegetation and having a long dry period." (Editora Nova Fronteira, 1987)

4. Interview with Dr. Carlos Magno on June 23, 2011.

5. Dr. Souza was born in South Rio Grande, a state famous for its soybeans, near the border with Uruguay. He started working at the agricultural research institute in the state of Bahia in 1973 and began to develop a tropical soybean variety, which was considered impossible at the time. In 1975, he moved to the CPAC, when it was located in the vicinity of the capital, Brasilia. Subsequently, he devoted his efforts to the breeding of a tropical soybean variety that could be cultivated immediately after the completion of the farmland reclamation in the Cerrado. In addition to the variety called Doko, which will be discussed later, he created about 30 varieties of tropical soybeans. "In my younger years, I learned a great deal from the approach of Yoichi Izumiyama [a Japanese crop cultivation researcher] among other researchers involved in the technological cooperation. Izumiyama then was like me today," he recalls. Dr. Souza came to Japan in 1991 on JICA's invitation and established ties with many Japanese researchers and research institutes. (This and other information regarding Dr. Pilinio Itamar M. de Souza is based on the authors' interviews with him on May 25, 2010 and January 12, 2011.)

6. It was named Doko after Toshio Doko, the former Chairman of Japan's leading business federation, Keidanren, who was committed to working for cooperation between Japan and Brazil.

7. This section owes much to papers by Dr. Eliseu Alves and the author's interviews with him in January 2011 and February 2012. See also Note 8.

8. Dr. Eliseu Alves was the Director in charge of technology development at EMBRAPA at the time of its foundation (1973). As its second President, he was largely responsible for laying the foundations of EMBRAPA as it is today. He has been referred to as the father of EMBRAPA, or "Mr. EMBRAPA." Now over 80 years old, he still holds an important post at EMBRAPA headquarters. Although Dr. Alves has not provided a strict definition of the EMBRAPA model, we can regard it as an excellent organization with several of the features he discusses, along with its activities (Chapter 6). Recently, a well known Brazilian agricultural magazine, *Globo Rural* (digital site), featured Dr. Alves in an article titled "Conversa com Eliseu Alves: Ele personifica a

Embrapa." The article emphasized that it is now a consensus in Brazil that EMBRAPA contributed to the agricultural revolution of the country. Since then, the development of a huge agro-industry has created a massive employment effect in Brazil (Globo Rural, 2014).

9. In the earliest years of the development of Cerrado agriculture, the majority of migrants in a pioneering program called PADAP were Japanese-Brazilians from the South and Southeast with experience in agricultural production and administration. These early farmers are discussed in Chapter 2.

10. For a discussion of the theories applied to EMBRAPA, this chapter drew on Fujita (2011) and Roll (2011).

11. For more information about CAMPO, see Chapters 2 and 10.

12. PADAP (Alto Paranaiba Guided Settlement Agricultural Program) will be discussed in Chapter 2.

13. People of Japanese descent accounted for 22 percent of the participants in PRODECER.

14. Project-type technological cooperation is one of the most comprehensive schemes of JICA's technical cooperation and partnership program.

15. For the technological achievements of the CPAC, see Chapter 7.

16. Dr. Norman E. Borlaug, who was made a Nobel Peace Prize laureate for his efforts in the Green Revolution approach, established the World Food Prize because there was no Nobel Prize for agriculture. Paulinelli and Lobato were the first Brazilian recipients of the prize.

## Bibliography

Alvim, Paulo de T. and Wilson de Araujo (1952) 'El suelo como factor ecológico en el desarrollo de la vegetación en el centro-oeste del Brasil'. *Turrialba*, 2(4), pp. 153–160.

Alves, Eliseu Roberto de Andrade, Geraldo da Silva e Souva, and Eliane Gonçalves Gomes (2013) 'Contribuição da EMBRAPA para Desnvolvimento da Agricultura no Brasil'. EMBRAPA, http://livraria.sct.embrapa.br/liv_resumos/pdf/00052960.pdf, date accessed May 21, 2014.

Alves, Eliseu Roberto de Andrade (2012a) 'Tecnologia e preservação são irmãs gêmeas'. *Agroanalysis*, 32(9), pp. 6–8.

Alves, Eliseu Roberto de Andrade (2012b) 'EMBRAPA: A successful case of institutional innovation' in G.B. Martha Jr. and J.B. de Souza Ferreira Filho (eds) *Brazilian Agriculture: Development and Changes* (Brasilia: EMBRAPA). (This chapter is a revised version of an article of the same title published in *Revista de Politica Agricola*, year XIX, special edition, July 2010, pp. 64–72.)

Bertran, Paulo (1994) *História da terra e do homem no planalto central: eco-história do distrito federal do indígena ao colonizador* (Brasilia: Solo Editores).

Correa, Pablo and Cristiane Schmidt (2014) *Public Research Organizations and Agricultural Development in Brazil: How Did EMBRAPA Get It Right?* Washington D.C.: World Bank.

Editora Nova Fronteira (1987) *Dicionário Aurélio* (Sao Paulo: Editora Nova Fronteira).

EMBRAPA (2000) *A cultura da soja no Brasil* (Londrina: EMBRAPA/CNPSo (CD Room)).

Faleiro, Fábio Gelape and E.S. Sousa (eds) (2007) 'Pesquisa, desenvolvimento e inovação para o Cerrado' in EMBRAPA (ed.) *Cerrados, Planaltina*, pp. 69–72. (Brasilia: EMBRAPA)

Faleiro, Fábio Gelape and Austeclinio Lopes de Farias Neto (2009) 'Savanas: demandas para pesquisa' in EMBRAPA (ed.) *Cerrados, Planaltina*, pp. 31–40. (Brasilia: EMBRAPA)

FAO and The World Bank (2009) *Awaking Africa's Sleeping Giant: Prospects for Commercial Agriculture in the Guinea Savannah Zone and Beyond* (Roma and Washington D.C.: FAO and The World Bank)

Ferri, Mário G. (1964) 'Informações sôbre a ecologia dos cerrados e sôbre a possibilidade de seu aproveitamento'. *Silvicultura*, 3(3).

Ferri, Mário G. (1971) 'Histórico dos trabalhos botanicos sobre o Cerrado' in Edgardo Blücher (ed.) *Simpósio sôbre o cerrado* (Sao Paulo:Editôra Italiana).

Ferri, Mário G. (ed.) (1977) *IV simpósio sobre o cerrado: bases para utilização agropecuária.* (Sao Paulo: Editôra Itatiaia).

Fujita, Yasuo (2011) 'What Makes the Local Government Engineering Department (LGED) So Effective?: Complementarity between LGED Capacity and Donor Capacity Development Support,' JICA Research Institute Working Paper (Tokyo: JICA).

Globo Rural (2014) 'Conversa com Eliseu Alves: Ele personifica a EMBRAPA,' April 9.

Goedert, J. Wenceslau (ed.) (1986) *Solos dos Cerrados: tecnologias e estratégias de manejo* (São Paulo: EMBRAPA-CPAC).

Gu, Shulin, John O. Adeoti, Ana Cecilia Castro, Jeffrey Orozco and Rafael Diaz (2012) 'The Agro-food Sector in Catching-up Countries: A Comparative Study of Four Cases' in F. Malerba and R. Nelson (eds) *Economic Development As a Learning Process: Variation Across Sectoral Systems* (Cheltenham, UK and Northampton, MA, USA: Edward Elgar).

Hayashi, Shigeru (ed.) (2008) *Centenário da imigração japonesa no Brasil e cinquentário da presença Nikkey em Brasília* (Brasilia: FEANBRA).

Imin 80-shunen-shi hensan iinkai [Committee for the 80 year history of Japanese immigrants in Brazil] (1991) 'Burajiru nihonimin 80 nenn-shi' [The 80-year history of Japanese immigrants in Brazil].(Sao Paulo: Imin 80-shunen-shi hensan iinkai)

Inter-American Development Bank (IDB) (2010) *The Age of Productivity: Transforming Economies from the Bottom Up* (Washington, D.C.: IDB).

Instituto Brasileiro de Geografia e Estatística (1979) *Região do cerrado: uma caracterização do desenvolvimento do espaço rural* (Rio de Janeiro: IBGE).

JICA (2006) *Relatório de avaliação final: projeto de conservação de ecossistemas do cerrado: corredor ecológico do cerrado Paraná-Pirineus* (Brazil: JICA Brazil).

Hongo, Yutaka (2002) Agricultural development in the Cerrado region of Brazil: the Japan–Brazil Agricultural Development Cooperation Program, perspectives and challenges. *Japan Journal of Tropical Agriculture*, 46(5), pp. 364–372.

Malerba, Franco and Richard Nelson (2012) *Economic Development As a Learning Process: Variation Across Sectoral Systems* (Cheltenham, UK and Northampton, MA, USA: Edward Elgar).

Marchetti, Delmar and Antonio Dantas Machado (eds) (1979) *Cerrado: uso e manejo: V Simpósio sobre o Cerrado: EMBRAPA-CPAC, CNPq.* (Brasilia Editerra).

Martius, C.F.V. (1951) A fisionomia do reino vegetal no Brasil. *Boletim geográfico*, 8(95), pp. 1294–1311.

Ministerio da Agricultura (1972) http://www.agricultura.gov.br/, date accessed May 23, 2014.

Ministry of Agriculture, Livestock and Supply (MAPA)(Brazil) and Japan International Cooperation Agency (JICA). (2002). *Japan–Brazil Agricultural Development Cooperation Programs in the Cerrado Region of Brazil: Joint Evaluation Study General Report.* Brasilia and Tokyo: MAPA and JICA

Miranda, Evaristo Eduardo (2013) *Agricultura no Brasil do Século XXI* (Brazil: Metalivros).

Mizumoto, Celso Norimitsu (ed.), Yoshisuke Ogura, and Julio Cesar Augusto Sesma da Cruz (2009) *O cerrado e o seu brilho* (Caramuru: Associação dos Antigos Escoteiros e Escotistas Caramuru).

Noman, Akbar and Joseph Stiglitz (2012) 'Strategy for African development' in Akbar Noman, Kwesi Botchwey, Howard Stein and Joseph E. Stiglitz (eds) *Good Growth and Governance in Africa: Rethinking Development Strategies* (Oxford: Oxford University Press).

Oliveira, Leandro dos Santos (2008) *Distribuição geográfica de espécie nativas do Cerrado: Resultados preliminares* (Brasilia EMBRAPA Cerrados).

Pessoa, José (2001) *Raízes de São Gotardo* São Gotardo: José Pessoa.

Roll, Michael (2011) 'The State that Works: 'Pockets of Effectiveness' as a Perspective on Stateness in Developing Countries,' Working Paper No. 128, Department of Anthropology and African Studies, Johannes Gutenberg University, http://www.ifeas.uni-mainz.de/Dateien/AP128.pdf, date accessed May 23, 2014.

Rydlewski, Carlos and Clarice Couto (2011) 'Por que a EMBRAPA é o nosso Google', 31 January, Época Negócios, Brasil.

Souza, Plinio Itamar Mello, Carlos Roberto Spehar, Claudete Teixeira Moreira, Romeu Afonso De Souza Kiihl, Leones Alves Almeida, Austeclínio Lopes De Farias Neto, Renato Fernando Amabile, Maurício Silva Assunção, Luis Claudio De Faria, Gottfried Urben Filho, and E. Pedro Manoel Figueira De Oliveira Monteiro (2000) 'Novas cultivares de soja para o sistema de produção de Grão no Cerrado'. *Pesquisa Agropecuária Brasileira*, 35(2), pp. 467–470.

Souza, Plinio Itamar Mello and Neylson Eustáquio Arantes (1993) *Cultura da soja nos cerrados* Brasilia: Associação Brasileira para Pesquisa da Potassa e do Fofato.

The Economist (2010). 'Brazilian agriculture: the miracle of Cerrado'. *The Economist*, August 26.

The World Food Prize Laureates (2006) http://www.worldfoodprize.org/en/laureates/20002009_laureates/2006_lobato_mcclung_paolinelli/. date accessed July 7, 2015.

Warming, Eugenius and Mário Guimarães Ferri (1973) *Lagoa santa e a vegetação dos Cerrados Brasileiros* (Belo Horizonte: Itatiaia; and São Paulo: EDUSP).

* In addition to the works listed above, this chapter draws on roughly 75 hours of interviews conducted by the authors with those who were involved in these programs.

# 2
# Establishment and Early Development: PRODECER Sets Agricultural Development in the Cerrado on Track

*Akio Hosono and Yutaka Hongo*

## Introduction

This chapter discusses the establishment and early development periods of Cerrado agriculture. These periods were preceded by the preparatory (inception) period, the major breakthrough of which was the Alto Paranaíba Guided Settlement Agricultural Program (in Portuguese, Programa de Assentamento Dirigido de Alto Paranaíba, PADAP). PADAP was promoted by the state of Minas Gerais (MG) in cooperation with the Cotia Agriculture Cooperative (hereafter Cotia), which was founded by Japanese–Brazilians in 1927 and grew to be the largest agricultural cooperative in the country. PADAP was the first structured program to demonstrate the feasibility of Cerrado agriculture and agri-business development. The starting point was São Gotardo in the state of MG.

On the basis of the successful PADAP experience, the Japanese–Brazilian Cooperation Program for Cerrados Development (PRODECER) was launched by the governments of the two countries to extend Cerrado agriculture to other areas of MG. The pilot projects of the first phase of PRODECER fully demonstrated the feasibility and high potential of Cerrado agriculture. The second phase of PRODECER carried out full-fledged projects in MG, as well as in Goiás (GO) and Mato Grosso do Sul (MS) states. At the same time, PRODECER also started pilot projects in the states of Bahia (BA) and Mato Grosso (MT). The third phase of PRODECER covered the Tocantins (TO) and Maranhão (MA) states. In this way, PRODECER was extended by its implementing and

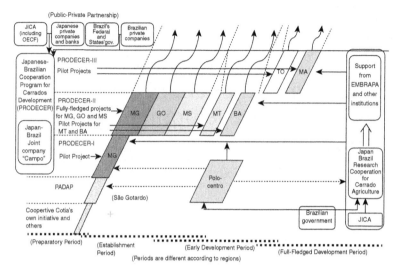

*Figure 2.1*   PADAP, PRODECER, and development of Cerrado agriculture
*Source:* Authors.

coordinating organization, CAMPO (Agricultural Promotion Company), from the core regions to the frontier regions of the Cerrado.

This chapter will analyze how innovative programs and institutions such as PADAP, PRODECER, and CAMPO set agricultural development of the Cerrado on track. Figure 2.1 shows PADAP, the scaling-up process of PRODECER, and the development of Cerrado agriculture. At the bottom of Figure 2.1, approximate timeframes of the four periods of Cerrado agriculture development are indicated. They are the preparatory, establishment, early development, and full-fledged development periods; their timeframes differ according to the region.

## 2.1   Pioneering initiatives for breakthrough in Cerrado agriculture: PADAP

When the governments of Brazil and Japan made the decision to officially promote Cerrado agriculture, PADAP served as their inspiration. As discussed in the following section, Cerrado agriculture was promoted through the Japanese–Brazilian Cooperation Program for Cerrados Development (PRODECER), which in turn was modeled on the PADAP in MG. The soybean and coffee cultivation of the early 1970s served as the background to PADAP's inception. This happened in Serra do Salitre,

MG, in a corner of the Cerrado plateau; Kazufumi Ogasawara was the Managing Director in charge of Cotia's business development. Because these cultivation efforts were very successful, Cotia was certain that there was great potential for agricultural development in the region; therefore, Ogasawara decided to lead efforts to develop the Cerrado. The area in question is near the so-called Minas Triangle (in Portuguese, Triângulo Mineiro), the part of MG closest to the state of São Paulo (SP). In 1971 and 1972, Ogasawara made his pioneering efforts toward development in the region (Cotia, 1987, pp. 158–159).

At approximately the same time, the MG state government, under Governor Rondon Pacheco, announced the policy of establishing agricultural development as the new pillar of the state's development. Through regional development projects, the state government was trying to promote agricultural development undertaken in collaboration with agricultural cooperatives. Cotia, headquartered in SP, was named by the state of MG as the strongest of these cooperatives. Cotia was one of the most successful agricultural cooperatives in Brazil; it was well known nationwide, and shipped the greatest volume of products to the fruit and vegetable markets of SP. In spite of Cotia being based in the states of SP and Paraná (PR), the MG state government asked for its cooperation. Professor Alysson Paulinelli, who had been appointed by Governor Pacheco as the MG Secretary of Agriculture, led the agricultural development in MG's Cerrado region.

The MG state government called on Cotia, and Cotia responded with the Chapadão development project in Alto Paranaíba under Ogasawara's leadership. However, this project differed greatly from what the state government had in mind: it was unprecedented and groundbreaking as an agricultural development project for the Cerrado, on a far greater scale than the state government had planned. In addition to the strong political will of the governor, Secretary Paulinelli was determined to promote the scheme, and Aluizio Fantini Valerio, President of the MG State Agricultural Corporation (Fundação Rural Mineira, also known as Ruralminas), also indicated his support. Thus, the MG state government as a whole decided to promote the Cotia project, which came to be called PADAP under Paulinelli's leadership.

Cotia's chairman, Gervásio Tadashi Inoue, visited the state capital of Belo Horizonte at the invitation of the state government; he met Governor Pacheco in April 1972, and they reached an agreement on PADAP. Governor Pacheco contacted President Garrastazu Medici, and a presidential order was issued in 1973, enabling the National Institute of Agrarian Reform and Settlement (INCRA) to buy land held by absentee landowners in order to implement PADAP. PADAP therefore constituted the first integral and structured project for establishing Cerrado agriculture. In its book entitled *History of 60 Years*, Cotia stated that "PADAP, initiated by Cotia, was the first full-fledged project for Cerrado development. PADAP, which attempted this new endeavor, propelled the nation's development. It was a distinguished plan worthy of mention in the cooperative's history of 60 years." (Cotia, 1987).

Of particular interest is how PADAP acquired the technologies that enabled grain to be cultivated in the barren land of the Cerrado. Numerous constraints inevitably stood in the way of achieving this unprecedented accomplishment. Why did members of Cotia, who were accustomed to the fertile *terra roxa* soil (alfisol) in SP and PR, decide to invest a large percentage of their resources and dedicate their lives to Cerrado agriculture in MG? What was behind their determination and confidence? The answer lies in the organizational power of Cotia, the passion of Ogasawara's leadership, the high educational level and knowledge of agriculture in southern Brazil among the second-generation Japanese–Brazilians who settled in the region, and above all their commitment to intensively work on the new challenge.[1]

Most of the young men who settled in the area were Japanese–Brazilians who belonged to Cotia. They shared a common drive and the beliefs of their parents, the first-generation settlers who had overcome numerous difficulties and dedicated their lives to agriculture in SP and PR. From their parents, the young Japanese–Brazilians had learned the principles of agricultural administration. Also common among them was the knowledge and technologies they had been exposed to at Brazil's most advanced agricultural educational institutes. Carlos Ogasawara, the first son of administrator Ogasawara of the northern PR branch of Cotia (in Londrina, PR), who led the project, had graduated from the Luiz de Queiroz College of Agriculture at the University of São Paulo and worked with PADAP. Isidoro Yamanaka (who later served on the technical staff of Minister of Agriculture Paulinelli), who is familiar with the details from that time, has said that many of the young second-generation Japanese–Brazilians were college/university graduates with a degree in agronomy (engenheiro agronomo).[2]

These graduates were able to learn from the experiences of Ogasawara, who had initiated test cultivation in the Cerrado. They were familiar with papers by Professor Mário Guimarães Ferri (see Chapter 1). They also learned from the research results of the Agronomical Institute of Campinas (IAC), which was at the forefront of soybean research, and from information offered by the IBEC Research Institute of the Rockefeller

Foundation. The IAC worked on the breeding and selection of soybean varieties. The soil of the site where IBEC is located in SP possesses the features of the Cerrado. The Agricultural Research Corporation of Minas Gerais State (EPAMIG), with its substantial budget, contributed significantly to technological development for Cerrado agriculture in MG. Many outstanding human resources were at the disposal of this institute, where research on the potential development of the Cerrado had begun. These resources included Dr. Toshiyuki Tanaka, a first-generation Japanese–Brazilian researcher.

It was thus an ideal time for testing the possibilities of agricultural development in the Cerrado. The collaboration between Governor Pacheco and Cotia in promoting PADAP presented an opportunity for success.

At the site in São Gotardo, MG, in the southeast corner of the Cerrado plateau, at altitudes from 1,150 m to 1,190 m, the second-generation Japanese–Brazilians utilized their experience in agricultural administration, university studies on features of the Cerrado, agricultural training in the U.S., and information available to them at the time. Essentially, the PADAP farmers were the most capable pioneers one could conceive of at that time. Most Brazilian soils are acidic, and the Cerrado soils in particular are very acidic and have very low fertility, but the settlers had knowledge of technologies to control soil acidity by limestone application.

In 1974, Cotia opened an office in the city of São Gotardo as the PADAP center, and declared the city to be the base for Cerrado development. The 24 first phase settlers were selected by April of the same year. These were all Cotia members from the states of SP and PR. The selection of 65 second phase settlers was finalized in December, and settlement began with a total of 89 farmers. Full-scale production began in 1979, when over two tons of soybeans per hectare were recorded, and 3.6 tons of wheat was harvested in irrigated plots. The potential for the cultivation of grain in the area had been demonstrated. President Ernesto Geisel visited PADAP, along with Minister of Agriculture Paulinelli, and praised the accomplishment of turning the Cerrado's barren land into an outstanding grain production ground.

In order to produce familiar Japanese food such as soy sauce, miso, tofu, and natto, Japanese–Brazilians had been enthusiastic about cultivating soybeans since the time that they immigrated. The thesis by Shiro Miyasaka (1958), who had earned his doctorate in agriculture at the IAC in Campinas, São Paulo, was titled *Breed Improvement of Soybeans*.[3]

Dr. Elizana Bardisella, who has studied contributions by Japanese and Dutch immigrants to Brazil in agricultural development, mentioned the cultivation of buckwheat, rice, soybeans, corn, and other grains as contributions made by the Japanese farmers, with an emphasis on breed improvement, technological improvement for cultivation and dissemination, and export of soybeans (Bardisella, 2004, p. 12).

## 2.2   An integral approach with financial and technological support to family farms: from PADAP to PRODECER

PADAP confirmed the feasibility of agricultural development by family farms in the Cerrado, but this did not mean that a stream of farmers and private-sector businesses would immediately start making inroads into the region. The Cerrado region is vast, with diverse natural, social, and economic environments. Numerous difficulties were foreseen for ordinary farmers or private-sector businesses that tried to develop their interests there. Cerrado development required many pioneer projects, as well as the establishment of agricultural technology systems and regional development models. These roles were to be performed by the government, and the Brazilian government was in need of international cooperation.

Japan and Brazil agreed to co-develop one of the largest-scale programs in the history of international cooperation, which would be based in the Cerrado area. The Brazilian government established an agricultural research institute in advance (EMBRAPA, as discussed in Chapter 1), organized agricultural financing systems, and improved the social and economic infrastructure (roads, the supply of electricity to rural villages, ports and harbors, and storage facilities). In 1975, in its efforts to prepare systems for Cerrado development, the Brazilian government formulated the Cerrado Development Program (POLOCENTRO, or Programa de Desenvolvimento dos Cerrados) as a national policy. As discussed in Chapter 1, EMBRAPA established the Cerrado Agricultural Research Center (CPAC) near the capital, Brasilia, in the same year.

Meanwhile, Japan stepped up its support via official development assistance. In 1977, JICA launched its first technical cooperation project with the CPAC. In addition to providing resources for the purpose of agricultural research in the Cerrado, Japan made a major contribution to the CPAC's research capabilities through human resource development initiatives, as discussed in the previous chapter. Tropical varieties of soybeans were bred, and technology was developed to improve soil fertility and management. Other breakthroughs included improvements in crop cultivation and in environmental preservation technologies.

In 1979, Japan began providing financial assistance for the full-scale promotion of agriculture in the Cerrado, via the Japanese–Brazilian Cooperation Program for Cerrados Development (PRODECER). Using PADAP as a template, the Japanese and Brazilian governments spent five years creating an innovative program management system. In an effort to expand the scope of frontier agriculture in the Cerrado, PRODECER created 21 settlements covering a total area of 345,000 ha (approximately 1.5 times the size of Tokyo) over a period of 22 years through agricultural cooperative-led schemes to develop family farmers' settlements. Taking the initiative, PRODECER set out to demonstrate the feasibility of agricultural development, and Japan provided both technical and financial cooperation during the initial stages of development in the region.[4]

In terms of institutional set up, PRODECER had three important components, the first of which was a financial cooperation mechanism. As is illustrated in Figure 2.3 (upper left to the middle), the Japanese government, through JICA and the Overseas Economic Cooperation Fund (OECF), and Japanese private banks provided finance to the Central Bank of Brazil (in later phases, to the Secretary of the National Treasury of the Ministry of Finance), which channeled the funds to farmers and cooperatives through Brazilian banks. As also illustrated in Figure 2.3, and as explained below, this process for providing finance to farmers was facilitated by CAMPO.

The second component was the Company of Agricultural Promotion (CAMPO, Compania de Promoção Agrícola), a bilateral public–private entity, established as an implementing and coordinating organization of PRODECER with a 51 percent share in participation from the Brazilian Agricultural Participation Company (BRASAGRO, Companhia Brasileira de Participação Agro-industrial) and 49 percent share from JADECO (Japan–Brazil Agricultural Development Corporation) (see the middle of the Figure 2.3; for more on CAMPO, see Chapter 10).

The third component was the collaboration of EMBRAPA and its various centers – especially the CPAC, for PRODECER. In short, PRODECER could be considered as an institution that facilitated financial and technological cooperation through its implementing and coordinating organization, CAMPO. The Brazilian government provided financing to PRODECER, established EMBRAPA, provided budgets for their research activities, and invested in BRASAGRO, which in turn invested in CAMPO. The Japanese government provided financial cooperation to PRODECER and technical cooperation to EMBRAPA, and invested in JADECO, which invested in turn in CAMPO.

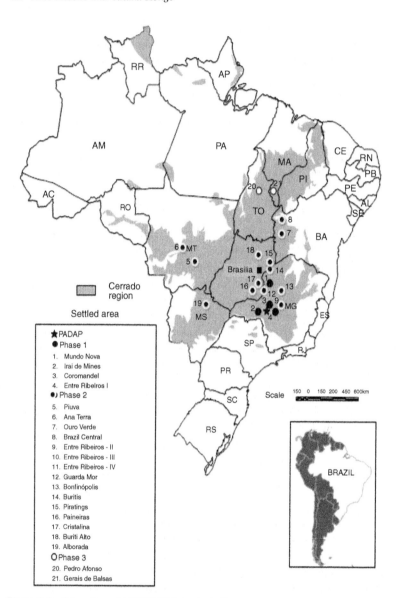

*Figure 2.2*   Cerrado and PRODECER project sites

*Source:* Authors, based on the definition of Cerrado according to the CPAC.

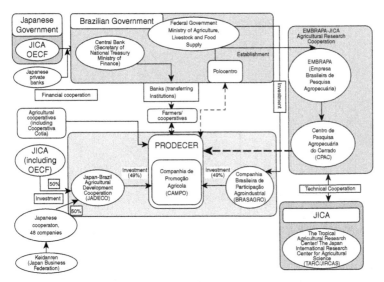

*Figure 2.3* Japanese–Brazilian Cooperation Program for Cerrados Development (PRODECER)

*Source:* Authors.

## 2.3 Farmers and organizations that played pioneering roles: PRODECER-I

From the start, it was mandatory for farmers participating in PRODECER to be Brazilian nationals. The successful PADAP effort with Cotia was used as a model, inspiring the Japanese side to send mainly Japanese–Brazilians as settlers in the early years of PRODECER, in order to take full advantage of their experience. Yet, there was also the opinion that MG farmers should be given priority, or that farmers should be recruited from all parts of Brazil, given that it was a national project. CAMPO, the Brazil–Japan public–private entity in charge of implementing PRODECER, studied this issue, and decided that of the total number of settlers, one-third should be Japanese–Brazilian farmers, one-third should be from MG, and the remaining third should be farmers from other states.

The settlement projects of the first phase of PRODECER (hereafter, PRODECER-I) were conducted in four locations, as shown in Figure 2.4.

The settlement site chosen by Japanese–Brazilian farmers was in the Paracatu region, which was also where most of the PRODECER-I sites

| 1972 | 77 | 1980 | 1985 | 1990 | 92 | 94 | 96 | 97 | 98 | 99 | 2000 | 01 |
|------|----|------|------|------|----|----|----|----|----|----|------|----|

| PRODECER I | PRODECER II | PRODECER III |
|------------|-------------|--------------|
| PRODECER : PHASE I (1977–1985) | PRODECER : PHASE II (1985–1992) | PRODECER : PHASE III (1995–2001) |
| (1) Project structure: pilot | (1) Project structure: pilot and full-fledged projects | (1) Project structure: pilot |
| (2) Project area: 60,000 ha | (2) Project area: pilot 60,000ha, full-fledged 140,000ha | (2) Project area: 80,000ha, |
| (3) Project site: 4 sites in one state (Farms settled: 92 Households) | (3) Project site: 15 sites in 5 states (Farms settled: 545 households) | (3) Project site: 2 sites in 2 states (Farms settled: 80 households) |
| (4) Amount invested: Japan: 5 billion 100 million yen Brazil: 5 billion 100 million yen | (4) Amount invested: Japan: 22 billion 100 million yen Brazil: 22 billion 100 million yen | (4) Amount invested: Japan: 7 billion 900 million yen Brazil: 6 billion 100 million yen |
| 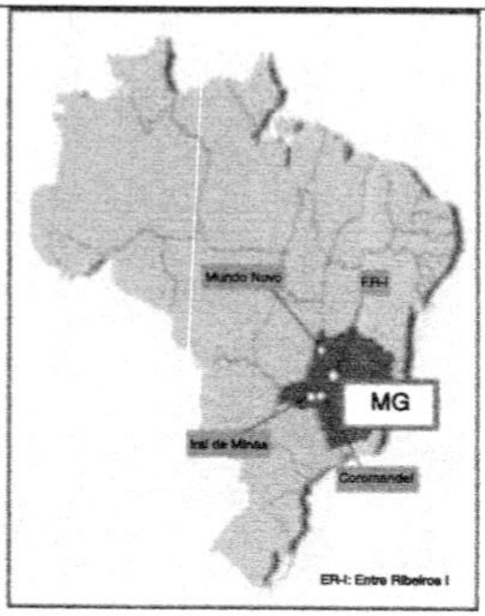 | 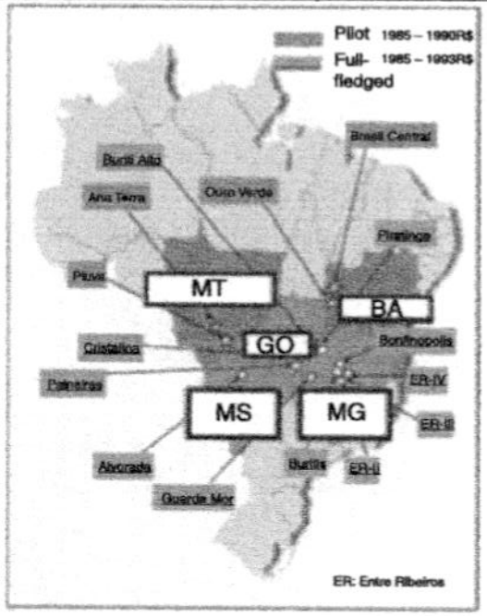 | 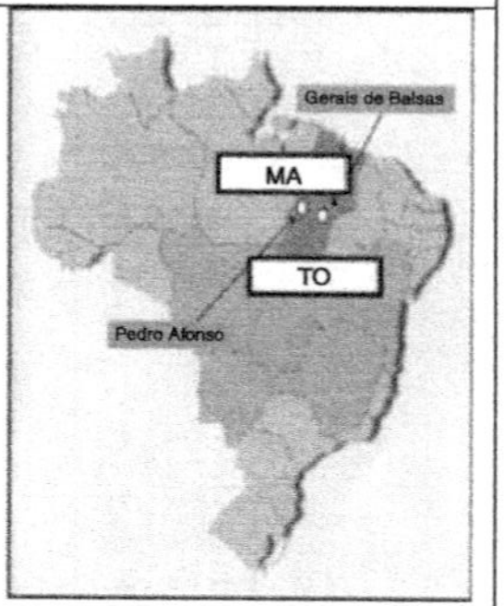 |
| Continuing process of confirming the feasibility of Cerrado agriculture. | The size of the project enlarged 5 times. On the other hand, Brazil's economy had slowed down and faced the hyperinflation. HQ of CAMPO was transferred to the capital, Brasília. | Project sites moved toward north. |

*Figure 2.4* Major features of the three phases of PRODECER

*Source:* Authors based on *Japan–Brazil Agricultural Development Cooperation Programs in the Cerrado Region of Brazil: Joint Evaluation Study General Report*, pp. 3–5; some adjustments have been made.

were concentrated. The area was later featured as the first case of the PRODECER project, and so the PRODECER settlement project in the Paracatu region is provided below as an example.

The land area of the project sites in the Paracatu region was 39,000 ha, and they consisted of Cotia's settlement project sites of Mundo Novo (meaning 'new world' in Portuguese; 23,000 ha) and those of two planting corporations (16,000 ha). PRODECER-I undoubtedly brought significant changes to the Paracatu region, changes that have been described by local people as revolutionary.

Until 1960, Paracatu was like an isolated island.[5] When Brazil's capital was moved from Rio de Janeiro to Brasilia in 1960, National Highway Route (BR 040) was paved between the state capital Belo Horizonte and Brasilia, and bridges were built across rivers. Paracatu, the capital city of this county, is about 230 kilometers southeast of Brasilia and situated along BR 040. However, it lacked even a single decent hotel and remained a somewhat sleepy village. The major industries were limited to extensive cattle-raising and dairy husbandry, as well as the cultivation of subsistence crops, using sedimentary soil along the river.

The county of Paracatu is located in the northeastern corner of MG, adjacent to GO to the west. It is located at 16 degrees south and 47 degrees west.[6] Before the project began, the population was only 22,000 (1977), and was mainly concentrated in the capital city of Paracatu and in small villages along the river.

PRODECER-I formally started on September 29, 1979. The Paracatu offices of CAMPO and Cotia were established in the city prior to the initial settlement. At the time, the local people responded coolly to the PRODECER project. They thought that agriculture would never be possible in the abandoned highlands of the Cerrado, and that the project itself was reckless. However, after development commenced in 1980 and Japanese–Brazilian settlers began to arrive, agricultural machinery and agri-business agents set up stores in town, resulting in intensive marketing activities and the invigoration of the entire community. The town's proximity to the capital city of Brasilia also invited media coverage. Gradually, the cataclysmic event that had taken place in the area was brought to the attention of the rest of the country.

Prior to the start of PRODECER, CAMPO began selecting the land lots for settlement. Through an analysis of aerial photos, CAMPO sought out plateaus suitable for mechanized farming. It then used twinjet planes flying at low altitude to survey the candidate lots and to visually confirm the topography and the condition of the rivers and roads. Agronomists from CAMPO also conducted on-site surveys on the vegetation, and analyzed soil samples to narrow down the candidate lots. Inspection of land registers and negotiations with landlords for land lot transactions were conducted through local real estate dealers, and specialized attorneys from CAMPO headquarters checked for errors in the documents and assessed their credibility. The Mundo Novo project land lots were purchased by February 1979 and zoning work for the settlement lots began at the site in early 1980. The areas along the river with abundant vegetation were designated as a sanctuary, or common reservation land (*condominio*). This *condominio* land has been preserved as a sanctuary to this day. Each zone was called a *lote* (lot), and all were set alongside a road and given access to the river whenever possible.

Cotia was the cooperative selected for Mundo Novo, and 48 farming families settled in the region, whose total land area was 23,000 ha. CAMPO took the initiative throughout the process, which was conducted democratically and fairly by what is now called the participatory approach. It was assumed that settlers would live on their lots, so they would not be absentee landlords; however, at times they were allowed to live in places away from the settlement, as far away as the country's capital (Paracatu), for reasons such as being able to send their children to school. Many settlers brought large trucks to their lots and slept on the loading platforms, or built huts to stay in during the agricultural land development period, supervising the development of the farm's land. Families with children below school age typically first lived in huts.

The majority of settlers were around 30 years old, with families typically comprising a husband and wife who were both tough and full of pioneering spirit. Most settlers were second-generation Japanese–Brazilians who had grown up seeing their immigrant parents overcoming numerous difficulties. CAMPO personnel recall their being determined and enthusiastic to become owners of mechanized farms, which their parents had not possessed.

Cotia established its office and facilities in Paracatu to support settlers, along with CAMPO. The agricultural land development was completed in September 1981. In 1982, construction work was finalized on a road stretching for 140 kilometers; 100 kilometers of electrical cables were also laid in the farming village. The total project cost was about 4.1 billion yen (US$20 million), excluding infrastructure, of which about 3.8 billion yen (US$19 million) was a loan from PRODECER, the remaining amount being borne by the settlers and Cotia. The scale of the PRODECER project was roughly 400 ha per family (reservation land included), with an investment amount of around 100 million yen (US$1.1 million).

The settlement projects in other areas proceeded similarly, although the settlers had different backgrounds. The land area of developed settlement sites in three locations (Paracatu, Irai de Minas, and Coromandel) reached 60,000 ha, with a total of 92 families.

In addition to the three settlement projects, another settlement project was implemented in 1983: Project Entre Ribeiros I, in the Santa Rosa region, Paracatu County. This project was conducted under the same rules, exclusively using resources from the Brazilian side and a property of 10,000 ha owned by CAMPO. This project benefited more than 41 families, and approximately US$17 million was invested.

Thereafter, the settler families were separated from CAMPO, and agricultural cooperatives were organized at each project site. They provided systematic support. This process took place as soon as settlers became independent. Within four years, PRODECER's most important goal was achieved: a level of productivity equal to that of the southern states, which was 2.2 tons per ha at the time, in comparison to the initial yield of one ton per ha in 1980. PRODECER-I thus proved to be a success, and demonstrated that Cerrado agriculture could be profitable.

## 2.4 Scaling-up to a full-fledged project: from PRODECER-I to PRODECER-II

In March 1981, a year and a half after PRODECER began, Japan and Brazil conducted an interim evaluation. The project received a positive assessment, in that it had progressed extremely well and had achieved a positive outcome. The success of PRODECER-I had groundbreaking significance, as it confirmed the profitability of the new industry called Cerrado agriculture. Having confirmed the success of PRODECER-I, the Brazilian side began to seek ways to expand the project, and to formulate specific plans for this purpose.

Cerrado agriculture had begun in MG, and Minister of Agriculture Paulinelli and his staff led Brazil's Cerrado development, as discussed above. MG was at the core of Cerrado development throughout the early years, and its state government established the infrastructure that encompassed all related institutes, which led to the success of PRODECER-I. However, after the evaluation, the federal government decided to relocate the CAMPO headquarters to Brasilia, on the basis that the Cerrado region extends also to other states of the Mid-Western part of the country. Relocation to the capital city turned out to be a big step forward, as PRODECER-II aimed to produce new development to promote Cerrado agriculture. Having its headquarters in

Brasilia strengthened the ties between CAMPO and the Ministry of Agriculture.

Preparatory work for PRODECER-II began when Japan's joint government and private-sector study team visited Brazil in March 1981. During the 1980s, Japanese financial cooperation was categorized into overseas investment and loan projects conducted by the Japan International Cooperation Agency (JICA), general projects of the Overseas Economic Cooperation Fund (OECF, later integrated with JICA after merging with JBIC), and projects funded by the Export-Import Bank. The loan conditions differed for each. The JICA investment and loan projects were intended for high-risk overseas pilot projects conducted by private-sector businesses. OECF funds were used for full-fledged projects, or when implementing a project on a larger-scale once it had demonstrated its viability in a pilot project. Export-Import Bank funds were allocated to low-risk, large-scale, private-sector investment and loan projects.

It was decided that for PRODECER-II, both JICA and OECF would participate in the following way. First, the PRODECER-I pilot project conducted in MG would be implemented on a larger-scale in areas of MG and other states with similar natural conditions. This project would be referred to as a full-fledged project.[7] Since the project risk was not very high, it would be treated as a general project funded by the OECF. Second, among the PRODECER-II projects, JICA would be in charge of projects with a higher-risk with regard to natural conditions. Such higher-risk projects were referred to as pilot projects, to be conducted in frontier areas of the Cerrado region outside of MG, where project feasibility had not been sufficiently demonstrated. In these areas, the natural conditions differed from those in MG.

Brazil's land area is 24 times that of Japan, and its biomes (biotic formations) are categorized into six groups: Amazon, Caatinga, Cerrado, Atlantic Rainforest, Campo, and Pantanal. Features found in the Cerrado are naturally diverse, as it occupies an extensive land area. What came into focus were the Cerrado frontiers adjacent to two biomes: the northeastern region of the Cerrado (the western region of BA, with annual rainfall of around 1,000 mm), adjacent to the Caatinga (a semi-arid region in northeastern Brazil); and the northern region of the Cerrado (in MT), which has higher rainfall (1,500 mm or more) and less fluctuation in daylight hours, which is adjacent to the Amazon (a high-temperature, humid area along the Amazon River). The Atlantic Rainforest area has a rolling landscape and is unfit for mechanized agriculture, and the Pantanal (a large wetland) is unfit for cultivation. The southern Campo (grasslands) is a traditional agricultural region of Rio Grande do Sul (RS), the southernmost state of Brazil, and is a long way from the Cerrado. Thus, the state of MT and the western region of BA were selected as the pilot project site.

The successful completion of PRODECER-I and reciprocal visits by the Japanese and Brazilian agriculture ministers completed the preparations for PRODECER-II. Also, at ground level, Japan dispatched several study teams to finalize the project content. When President João Baptista Figueiredo of Brazil visited Japan in 1984, he and Prime Minister Yasuhiro Nakasone jointly announced their intention to cooperate on PRODECER-II. The major items agreed on were pilot projects on a total of 50,000 ha of land in BA and MT, and full-fledged projects in the states of MG, GO, and MS. This officially established the content of PRODECER-II.

The preparation for PRODECER-II took five years. The Brazilian side first indicated its interest in PRODECER-II to Japan's Foreign Minister Sunao Sonoda in August 1981; five years later, in March 1985, the signing ceremony for PRODECER-II (the project contract and loan agreement) was held in Japan, officially marking the start of the project.

Unlike PRODECER-I, no planting companies participated in PRODECER-II; only agricultural cooperative-led family farm settlement development projects were conducted. Yet PRODECER-II was much larger in scale, consisting of projects with about five times the number of settlers, on land areas five times larger, in five states at the same time. (Some of the full-fledged projects were initially delayed at some project sites.) There were 15 settlement project sites, 545 settler families, and a total project land area of 205,000 ha (almost equal to the land area of Tokyo). The project lasted for eight years, from 1985 to 1993 (the pilot projects were completed by 1990),[8] with a total loan amount of 44.2 billion yen (US$375 million) by Japan and Brazil (Figure 2.4).

## 2.5 Coping with macro-economic difficulties: PRODECER-II

Even though the settlement method of family farms was the same as in PRODECER-I, a host of totally different and unexpected difficulties occurred during PRODECER-II, due to the uncertainties in the country's macro-economic environment at that time. PRODECER-II was conducted during a period of unprecedented turmoil in the Brazilian economy, as shown in Figure 2.5. At the beginning of 1985, inflation was running at 220 percent, and a moratorium on Brazil's debt repayment was declared in 1987. The inflation rate continued to surge, reaching 1,600 percent

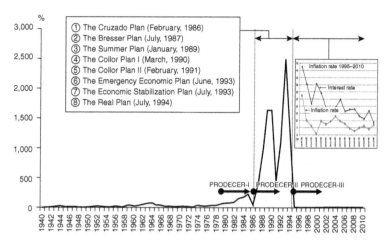

*Figure 2.5*   Evolution of consumer price index in Brazil
*source*: Compiled by authors based on FIPE and Central Bank statistics.

in 1990, and, following a temporary drop, climbed to a record high of 2,490 percent in 1993, when the project was completed. During this period, unorthodox (frequently called heterodox) economic stabilization measures were implemented seven times, all of which failed, and the entire Brazilian economy plunged into chaos. Along with other Latin American nations where inflation had intensified, Brazil first implemented orthodox measures based on monetarism, including the adjustment of interest rates and currency supplies, but none of these approaches worked. Thus, the countries implemented further measures to suppress inflation, such as price freezing and income policy.[9] Under these circumstances, agricultural policies inevitably faced substantial change.

For example, Brazil conducted re-denomination four times in the period between the Cruzado Plan in February 1986 and the start of the Real Plan at the end of June 1994. During these eight years and four months, the value of the Brazilian currency fell to 1/2,750,000,000,000. At supermarkets, 'human wave tactics' were adopted to replace price tags, usually first thing every morning. Prices for the same products by the same manufacturers differed by 40 percent or 50 percent at adjacent stores, and the difference reached as high as 420 percent for some products (Suzuki, 2002, pp. 21–22). Loan principal swelled by indexation (correction of currency value), and was then subject to the world's highest interest rates. On a given date, balance sheets from three months before no longer represented the actual situation. Businesses went bankrupt at a rapid speed (Suzuki, 2001, pp. 23–24). This was the economic situation that agricultural producers in the Cerrado faced.

**How did PRODECER overcome the macro-economic difficulties?**

An inflation correction ratio was adopted for interest rates in 1985; the value of debts was also corrected, causing them to swell, and haphazard agricultural loan conditions were changed six times. The minimum prices for agricultural products were lowered after 1987, in an attempt to suppress inflation. Agriculture in Brazil suffered a tremendous overall blow due to the gap that formed between hyperinflation and the high interest rate policy on the one hand, and the suppression of the correction rate for the prices of agricultural products on the other. Farming families fell into default and agricultural cooperatives went bankrupt one after another. Supermarkets could take the time to replace price tags overnight, but restrictions on product prices were imposed on agricultural producers by the government. Cotia, which was the largest agricultural cooperative, and had always been regarded as a model for Brazil, fell into default in 1993 and was forced to dissolve in 1994. Almost 30 percent of Brazil's agricultural cooperatives went bankrupt during this period (Tajiri, 1999). Normalization of the chaotic Brazilian economy finally began when the Real Plan[10] was introduced in July 1994.

Remedial measures were implemented for heavily indebted farmers. These included the Securitização,[11] introduced in 1996, which mitigated the burden for farmers with debts of 200,000 reais (equivalent to about US$200,000 at the time) or less, and PESA,[12] introduced in 1998, which provided relief for debts of more than 200,000 reais, which were not covered by the Securitização. These systems have been continuously revised and improved upon to this day. Many farmers avoided compulsory seizure by banks and further benefited from the remedial measures.

PRODECER was also affected during this period by delays to road construction and electrification work in rural villages due to the restricted financial conditions of the state governments.

In spite of the historically adverse macro-economic conditions of Brazil, the agricultural development projects of PRODECER-II were completed in a land area 5.5 times of PRODECER-I. Hence the results and management capabilities of PRODECER-II are highly noteworthy. According to surveys conducted separately by CAMPO in 1997 and OECF in 1999, the withdrawal rate among settlers was 23 percent for the pilot projects and 27 percent for the full-fledged projects. Therefore, over 70 percent of the farmers survived the very difficult economic conditions. The relatively low ratio of PRODECER settler farmers who withdrew despite the great economic hardship can be attributed to the following factors. First, CAMPO negotiated measures on behalf of the settlers to mitigate debts with financial institutions (banks); state development banks usually agreed to loan reductions and various remedial measures because they considered that PRODECER had contributed to agricultural development, although the Banco do Brasil, the largest bank in Brazil, remained inflexible and resorted to coercive measures such as compulsory seizures. Second, PRODECER had ample funds in store. Its projects were never abandoned due to an interruption in funding.

### PRODECER-II compared with PRODECER-I

Despite initial optimism for PRODECER-II, the outlook became grim when the project actually began. This was due to the economic crisis that hit Brazil and the economic policies that resulted. The conditions for carrying out PRODECER-II turned out to be the opposite of those for PRODECER-I.

Specifically, the high interest rate policy was implemented in order to suppress inflation, and a measure locally referred to as monetary correction was conducted, in which outstanding loans were adjusted to suit the inflation rate (a policy to change the amount of principal to suit the inflation rate). During PRODECER-I, the loan principal was never corrected. A preferential interest rate was also applied, which in most cases was lower than the inflation rate. In short, through its policies and financial measures, the national government was in effect supporting the pilot projects of PRODECER-I to establish Cerrado agriculture as an agri-business.

In the difficult period of business establishment, entrepreneurs (settler farmers) are highly likely to face difficulties they have never experienced before. They are therefore under conditions of high-risk, and support like that provided by PRODECER-I is essential. In the early development period, which comes after the feasibility of a business has been verified (which was accomplished during PRODECER-I), the need for preferential treatment gradually declines. During PRODECER-II, the government implemented measures that were opposite to those it had practiced during PRODECER-I. A high interest rate was applied instead of preferential interest rates, on top of which the value of the principal was corrected. These measures were not intended to target agriculture: they were implemented equally for all industries in Brazil. However, it cannot be denied that these policies had a particularly strong impact on agriculture. Unlike manufacturing industries and service industries, agriculture requires a long time from planting to harvesting, and therefore has difficulty adapting to rapid inflation and high interest rates. Preferential treatment was also eliminated. This completely changed the economic environment of Cerrado agriculture. It was never easy for the farmers to adapt to these rapid changes.

CAMPO and agricultural cooperatives expended the utmost effort to support settler farmers who ran into unanticipated difficulties. The efforts CAMPO made to reduce the burdens on settler farmers and in rescheduling the loan repayment period were extraordinary.

## 2.6 New frontiers addressing new challenges: PRODECER-III

The Brazilian government unofficially contacted Japan in 1988 about the start of the third phase of PRODECER (hereafter PRODECER-III), attesting that PRODECER-II had brought about significant agricultural development despite the adverse economic situation, and noting that requests had been made by the states of TO, MA, and Piauí (PI), which had not yet begun agricultural development in the Cerrado, to start similar development projects. Minister of Agriculture Iris Rezende Machado filed an official request to Japan in 1989. The amount of foreign currency flowing into Brazil had fallen drastically during the 1980s due to the accumulated debt crisis, and was reduced to nothing following Brazil's declaration of a debt moratorium in 1987. As such, demands for funding were high in Brazil. Study teams from Japan were dispatched to Brazil several times in response to this request, and senior officials from the Brazilian government visited Japan. Negotiations led to an agreement for Japan to provide assistance only to the pilot projects (see Note 3). The records of discussion (R/D) were concluded in 1993, and the project contract and loan agreement for PRODECER-III were signed at the Ministry of Agriculture of Brazil on March 29, 1994.

During the negotiation process, one cautious opinion suggested that Japan should not respond to the request for cooperation on this large-scale agricultural development project while Brazil was going through unprecedented economic hardship. The final decision, however, was that supporting PRODECER-III would help Brazil to suppress inflation and increase exports, and that it would ultimately stabilize the world food supply. Incidentally, five years were needed for the preparation of PRODECER-III, from the initial request to the conclusion of the project agreement.

In the end, 60 percent of the funding came from Japan, 30 percent from the Brazilian government, and 10 percent from the settlers and associations. The project was conducted in two sites in two states, and involved 80 farming families, who settled on a total land area of 80,000 ha. The total investment amount came to around 14 billion yen (US$138 million). The average land area allocated to each farming family was 1,000 ha, of which 50 percent (500 ha) was designated as a statutory reserve (protected land for environmental conservation).

The project sites were in the northernmost region of the Cerrado (at a latitude of 9° south), at low altitudes (240 meters for the Pedro Alfonso, TO, site and 540 meters for the Balsas, MA, site), and thus the annual average temperature was high, at between 25 °C and 26 °C. Annual rainfall was 1,200 mm at Pedro Alfonso and 1,590 mm at Balsas. Large irrigation facilities (center-pivot facilities) were installed in some areas to reduce the risk involved in agricultural operations, and crops such as pineapples, bananas, and coffee were introduced in addition to grains to establish a model of diversified agricultural operations.

PRODECER-III began in MA in July 1995 and in TO in July 1996. The start was delayed in TO, as the Banco do Brasil, which acted as the financing agent, requested a 50 percent loan guarantee from the TO state government. According to the Bank, the project risk was high, and time was needed for these negotiations. The TO state government eventually agreed to provide the loan guarantee. The Banco do Nordeste, which acted as the loan agent for MA, did not request such a guarantee from the state government.

PRODECER-III confirmed the feasibility of Cerrado agriculture in those frontiers located far from MG, the pioneer state of the agricultural development of the Cerrado.

## 2.7   Accomplishments of PRODECER

The accomplishments of the PRODECER can be summarized in four points:

### (1)   Dissemination of innovative technologies

The production technology system adopted at PRODECER project sites introduced the latest technologies using a participatory approach (including agricultural producers, research institutes such as EMBRAPA, loan institutions, and administrative institutions). Through this process, the people involved in the project learned from each other and shared their wisdom, as well as new technologies and knowledge. This process was maintained, and improved upon through continuous learning process. The basic system was family farms' mechanized, rain-fed upland farming.

The technologically innovative nature of PRODECER is evident in the numerous production technology systems introduced at the individual project sites in the wide and diverse area of the tropical savanna (the Cerrado), when the unique aspects of each site are taken into consideration. PRODECER simultaneously achieved an expanded land area and increased yields (productivity per unit of area).

Another factor behind the project's success is the incremental approach to scale up carefully. To start with, soybean cultivation was at the core of the production technology systems established for the farmers. PRODECER's influence reached far across the Cerrado region from its core project sites, and in this way innovations in technologies and institutions expanded in scale.

### (2)  Introduction and expansion of the system of agricultural cooperative-led guided settlement projects

PRODECER succeeded in developing a guided site settlement model that started with PADAP. Land lots were purchased from normally absentee owners of large-scale, currently unused land through local real estate dealers, and these lots were then distributed to medium-scale farming families (managed by families endowed with agricultural entrepreneurship), enabling more intensive land use. This could be described as a model of dispute-free, large-scale agrarian reform.

The process was also a socially inclusive development scheme, which earmarked land lots for settlement by local residents and promoted participation by regional communities. The project was never forced on a site from the outside. Some residents in regional communities were motivated to begin agricultural operations by PRODECER's example.

### (3)  Successful coordination in PRODECER: roles performed by CAMPO

As the core organization of PRODECER, CAMPO, a Japanese–Brazilian public–private joint venture, planned the projects, made adjustments, secured land lots for settlement, selected settlement cooperatives, selected settler farmers, and provided technological instruction, among other tasks. Even after the project's completion, CAMPO provided follow-up measures to keep the momentum of agricultural development at project sites and advice to cooperatives. Although CAMPO was engaged in public activities, it was capable of dealing with various issues at a speed comparable to that of a private-sector business. Chapter 10 of this book will discuss these outstanding characteristics of CAMPO.

Even ten years after the completion of PRODECER, CAMPO remains active in a range of areas, such as biotechnology, soil analysis, precision agriculture, and project planning (including consultancy work for neighboring countries in Latin America and on the tropical savanna in Africa). CAMPO's assets as a Japanese–Brazilian joint venture,[13] with regard to the experience of Cerrado agriculture, are very valuable.

As had been mentioned, Brazil fell into a debt crisis during the 1980s due to the rapid deterioration of its foreign currency balance. The economic turmoil at the time affected farmers nationwide, forcing many individuals and cooperatives into bankruptcy. Most of those affected were settler farmers who had relied on loans for the large initial investment. Some of them withdrew, as they ultimately became tied up by an agricultural administration and were unable to repay their loans. According to a survey conducted in the latter half of the 1990s, after the inflation period ended, the overall ratio of farmers who gave up farming (withdrew) was 35 percent for the first to third phases of PRODECER; these farmers sold their land lots and moved out. However, the lots did not turn into abandoned farmland. Instead, they were purchased by other farmers in the same settlement, or were cultivated by farmers who moved from other areas. The ratio of farmers who withdrew differed depending on the project period and site. Ironically, the number of farmers who withdrew was greatest in the first phase (PRODECER-I), which had offered the most favorable loan conditions, for the following reasons: (1) some farmers purchased large-scale irrigation facilities with loans acquired for the purpose of realizing year-round cultivation during the period of economic difficulties; and (2) development drove up the asset value of farming land lots, inspiring some farmers to sell their land lots and purchase larger-scale lots at low prices in areas deeper inland in order to expand the scale of their agricultural land. The latter practice is frequently seen in the frontier agricultural zones.

In any case, at the time of the survey at least 65 percent of the farmers had survived through the period of unprecedented economic problems. This success indicates that the farmers who participated in PRODECER had a strong entrepreneurial spirit and were highly capable.

## 2.8   Concluding remarks

As was discussed in Section 2.2 of this chapter, PRODECER inherited the main framework of PADAP: the agricultural cooperative-led guided settlement approach. However, the following differences between the two should be emphasized with regard to innovative institutional and technological development. First, PADAP was a project led by the MG state government, other state government agencies, and Cotia, while CAMPO, the Japanese–Brazilian joint venture, led PRODECER. Project coordination by CAMPO established 21 project sites (development bases) in seven states. With this type of large-scale agricultural development project, which was conducted almost nationwide, inflexibility could be expected if it was government-run; the nature of the project could greatly change depending on the policies of political leaders. The major innovative aspect of PRODECER is the fact that a public and private collaborative system involving the national, state, and county governments and the private sector had been introduced as a preliminary step for farmers and private businesses to independently initiate vigorous business development.

Differences can also be identified in terms of the operation scale and cultivation technologies. In PRODECER, agricultural technologies that rapidly developed after PADAP were incorporated in ways that were suited to each region: in other words, new technologies were continuously introduced. As a result, the land area by each farming family was 400 to 1,000 ha in PRODECER, from around 200 ha in PADAP. PRODECER verified the feasibility and potential of Cerrado agriculture and opened a path for ordinary farmers and corporate agricultural producers to participate in the new industry. The pilot projects of PRODECER-I were indeed successful, and clearly proved that the new Cerrado agriculture was profitable at the center of the Cerrado area of MG. In other words, it was convincingly demonstrated that Cerrado agriculture was possible as a business.[14] PRODECER-II and PRODECER-III were significant in that they scaled up Cerrado agriculture. The full-fledged projects in MG, the birthplace of Cerrado agriculture, and in its neighbor states of GO and MS, as well as pilot projects in BA, MT, TO, and MA were largely responsible for disseminating Cerrado agriculture to even wider areas and bringing it to the development period.

## Notes

1. For Japanese–Brazilians who settled in the Cerrado, see Sasaki (2008) and Mizumoto (2009).
2. Interview [May 18,2010] with Isidoro Yamanaka, who is familiar with the details from that time as he was a staffmember of Cotia Cooperative during 1950–59 and who later served on the technical staff of Minister of Agriculture Paulinelli.

3. For this study as well as the contribution of Japanese-Brazilians to agriculture in Brazil, see Yamanaka (2008).
4. A recent article by the Minister of Agriculture, Livestock, and Supply of Brazil, Katia Abreu, highlighted this aspect: "The contribution of Japanese cooperation was not limited to promotion of technological innovations, such as soil improvement and breeding, but was extended to institutional innovations, including development of land use in the form of cooperatives, which was fundamental for Cerrado development." (Abreu, Katia 2015; Translation from Protuguese bv authors) http://www.agricultura.gov.br/internacional/noticias/2015/07/contribuicao-da-soja-no-pib-brasileiro-e-seu-vinculo-com-o-japao, date accessed June 7, 2015.
5. The Paracatu region first appeared in the written history of Brazil when alluvial gold was discovered in the mid-18th century. The Paracatu River flows about 50 kilometers southeast of the municipality's capital city of Paracatu. The municipality's name derives from the river: 'Paracatu' is a native Tupi word meaning 'good river.' Paracatu once flourished in the deep inland area of Minas Gerais, but disappeared from history when the alluvial gold was exhausted. Even during the 1940s, it took six days to travel from the state capital Belo Horizonte (about 400 kilometers away) by horse, which was the only available means of transportation.
6. The location of Paracatu is equivalent to Manila in the Philippines at about 15 degrees north in the northern hemisphere, with an extensive land area of 8,232 square kilometers (3.7 times that of Tokyo's 2,187 square kilometers). There are 853 counties in MG; the land area of Paracatu ranks third in the whole of Brazil. The altitudes and topography of this county are roughly divided into three tiers: (1) the highland plateau, at 1,000–1,400 meters, where the Mundo Novo settlement was developed; (2) the central plateau, at about 700 meters, where the city of Paracatu is located; and (3) the lowland, at about 400–500 meters. It is particularly hot in the lowland area, with a climate the locals refer to as a natural greenhouse. Precipitation tends to be less at the lower altitudes.
7. For Brazil, there was no difference in the loan conditions between the pilot and full-fledged projects, so the categories had no significance, but Brazil went along with Japan in the categorization.
8. In 1989, OECF funded the PRODECER-III project at the request of the Brazilian government with the surplus funds from PRODECER-II; as a result, PRODECER-II was extended to 1993.
9. The first to be implemented was the Cruzado Plan, followed by the Bresser Plan (1987) and the Summer Plan (1989), all of which froze consumer prices and wages. In the Summer Plan, short-term national bonds at a high interest rate were issued. Yet all of them ended in failure, since they were able only to suppress inflation for a short time. The Real Plan implemented in 1994 finally suppressed inflation. For about four years following the failure of the Summer Plan, until the start of the Real Plan, even greater confusion reigned. The Collor Plan (first phase) resorted to the drastic and unprecedented measure of freezing not only consumer prices, but also bank deposits. This disabled the withdrawal of bank deposits above a certain amount.

10. The Real Plan introduced by Minister of Finance Fernando Henrique Cardoso (who later became Brazil's president) consisted of the following measures: (1) linking the new currency, the *real*, to the US dollar; (2) suppressing consumption and raising the reserve deposit rate by a high interest rate policy; (3) promoting market liberalization; (4) fixing public fees; (5) establishing a national bond redemption fund; and (6) introducing free negotiation concerning wages.

11. *Securitização* was a system which put a hold on repayments for ten years (during which time the debt was kept unchanged for two years), and an annual rate of 3 percent plus the fluctuation ratio of prices of agricultural products was applied as the interest rate.

12. PESA was a system in which the debtor purchased 20-year national bonds equivalent to the debt amount, and the debt would be canceled if the debtor paid the interest. The interest rate was set to the rate of inflation plus 8 percent.

13. Since JADECO, an investor company on the Japanese side, was liquidated in November 2007, Japan's investment ratio has declined from 49 percent to 9.96 percent (shares owned by two trading companies based in Japan).

14. Other projects which developed settlements in other parts of the Cerrado area at about the same time were not so successful. These projectss include the Vale do Gorutuba in northern MG, Lucas do Rio Verde in mid- and northern MT (PER LRV), and Coopermosa Barro Preto Cortel (Coopermosa) in western BA. In all, 93 percent of farmers withdrew from PER LRV, and 67 percent from Coopermosa. In contrast, the overall withdrawal rate was only 35 percent for PRODECER.

# Bibliography

Abreu, Katia (2015) 'Contribucao da soja no PIB brasileiro e seu vinculo com o Japao' (July 6th 2015) http://www.agricultura.gov.br/internacional/noticias/2015/07/contribuicao-da-soja-no-pib-brasileiro-e-seu-vinculo-com-o-japao, date accessed July 7, 2015.

Bardisella, Elizana (2004) 'Japanese and Dutch in the Modern Agricultural Development in Brazil,' Thesis presented to the Tokyo University of Agriculture and Technology.

Cotia-sanso-chuokai (1987) 'Cotia-sangyo-kumiai-chuokai 60-nen no Ayumi' [History of 60 years: Cotia Agricultural Cooperative]. Sao Paulo: Cooperativa Cotia.

D'Araujo, Maria Celina and Celso Castro (eds) (1997) *Ernesto Geisel* (Brazil: Editora Fundação Getúlio Vargas).

Faleiro, Fábio Gelape and Austeclinio Lopes de Farias Neto (2009) *Savanas: demandas para pesquisa'* (Planaltina: EMBRAPA Cerrados).

Faleiro, G.F. and E.S. Sousa (eds) (2007) *Pesquisa, desenvolvimento e inovação para o Cerrado* (Planaltina: EMBRAPA Cerrados).

Hayashi, Shigeru (ed.) (2008) *Centenário da imigração japonesa no Brasil e cinquentário da presença Nikkey em Brasília* (Brasil: FEANBRA).

Hongo, Yutaka (2002) 'Agricultural development in the Cerrado region of Brazil: the Japan–Brazil Agricultural Development Cooperation Program, perspectives and challenges.' *Japan Journal of Tropical Agriculture*, 46(5), pp. 364–372.

Imin 80-shunen-shi hensan iinkai (1991) 'Burajiru nihonimin 80 nenn-shi' [The 80-year history of Japanese immigrants in Brazil].Sao Paulo: Imin 80-shunenshi hensan iinkai.

Incêncio, Maria Erlan (2010) 'O prodecer e as tramas do poder na territorização do capital no Cerrado,' Thesis, Universidade Federal de Goiás, http://projetos.extras.ufg.br/posgeo/index.php/trabalho/o-prodecer-e-as-tramas-do-poder-na-territorializacao-do-capital-no-cerrado/, date accessed May 23, 2014.

Instituto Brasileiro de Geografia e Estatística (1979) *Região do cerrado: uma caracterização do desenvolvimento do espaço rural* (Rio de Janeiro: IBGE).

Japan–Brazil Agricultural Development Corporation, JADECO (2007) 'Nippaku Nogyokaihatsu-kabushikigaisha-shi :Serado Nogyo Kaihatsu kyouryoku-jigyou, 30-nen no Kiroku' [History of Japan–Brazil Agricultural Development Corporation: 30 years of the Japan–Brazil Cooperation Program for the Agricultural Development of the Cerrado].: Tokyo: JADECO

JICA (1979) *Regional Development Study of the Three States: Esprito Santo, Minas Gerais, and Goias, Federative Republic of Brazil* (Final Report) (Tokyo: JICA).

JICA (1980) *Japan–Brazil Agricultural Research Cooperation project reference materials* (Tokyo: JICA).

JICA (1989) *Evaluation Report on the Second Pilot Project of Japan–Brazil Cooperation Program for the Agricultural Development of the Cerrado Region* (Tokyo: JICA).

JICA (2002) *Japan-Brazil Agricultural Development Cooperation Programs in the Cerrado Region of Brazil: Joint Evaluation Study: General Report* (Tokyo: JICA).

Miyasaka, Shiro (1958) 'Contribuição para o melhoramento da soja no estado de São Paulo,' Thesis, E. S. A. 'Luiz de Queiro,' U. S. P. Piracicaba.

Mizumoto, Celso Norimitsu (ed.) (2009) *O cerrado e o seu brilho* (Caramuru: Associação dos Antigos Escoteiros e Escotistas Caramuru).

Pessoa, José (1999) *São Gotardo: Sua gente sua evolução* (Belo Horizonte: Editora Lutador).

Pessoa, José (2001) *Raízes de São Gotardo* (Brasil: São Gotardo).

Santo, Benedito Rosa do Espirito (2001) *Os caminhosnda agricultura brasileira* (Brasilia: BM&F Brasil).

Sasaki, Luiz Isamu (2008) *Portal do Cerrado* (Bela Horizonte: Editora O Lutador).

Suzuki, Takanori (2002) *Brazil no chosen: Sekai no seicho senta o mezashite* [Challenge of Brazil: Toward a world center of growth] (Tokyo: Japan External Trade Organization).

Tajiri, Tetsuya (1999) *Brazil shakai no rekishi monogatari* [A history of Brazilian society] (Tokyo: Japan Marketing Education Center).

Yamanaka, Ishidoro (2008) 'Influência do Imigrante Japonês na Agricultura Brasileira' in Kiyoshi Harada (ed.), *O Nikkei no Brasil* (São Paulo: Cadaris Comunicacão).

# 3
# Development of Cerrado Agriculture: The Path to Becoming a Major Global Breadbasket

*Akio Hosono and Yutaka Hongo*

## Introduction

This chapter discusses the autonomous development process of Cerrado agriculture once its feasibility and potential had been proven in the three phases of PRODECER (Japanese–Brazilian Cooperation Program for Cerrados Development) in their respective regions. First, farmers from southern states began to move into the Cerrado. Simultaneously, a few transnational grain trade companies (GTCs) (or 'grain majors') rushed to expand their business in many areas. Section 3.1 of this chapter provides an overview of this dynamic process. The extraordinary increase in production of soybeans and corn in the Cerrado is analyzed in Section 3.2. This process was accompanied by the development of transportation and logistics systems (Section 3.3). Transnational GTCs were one of the drivers of the accelerated expansion of Cerrado agriculture in its autonomous development phase (Section 3.4). At the same time, Cerrado agriculture increasingly diversified: today, the Cerrado produces not only grains, but also a wide range of agricultural products such as coffee, cotton, sugarcane, vegetables, and fruit. Moreover, livestock is raised in the region, and processed foods are also produced (Section 3.5).

## 3.1 Northward expansion of the frontier of Cerrado agriculture

The world's grain-growing regions have been utilized for much of human history. The most important determining factor in grain production is the condition of the natural environment, which defines the types of grain that can be cultivated. Soybean, for instance, is grown in temperate regions with varying hours of sunlight and ample rainfall, while wheat is grown in cool climates without much rainfall during the harvest period.

However, breeding new crop varieties and developing cultivation technologies specifically for the Cerrado region overcame these limitations in the natural environment, quickly transforming the region into one of the world's major breadbaskets.

The pioneering crop that led agricultural development in the Cerrado was soybean. After the Cerrado area had been confirmed as suitable for grain cultivation, the Cerrado agricultural frontier began to shift northward. As the frontier in the United States shifted from east to west during the 1860s, the development front in Brazil shifted from the temperate south to the tropical north (the Cerrado) during the 1970s, a century later. In just a quarter of a century, this frontier reached the Cerrado region in Roraima (RR), situated in the northern hemisphere across the Amazon. A 'big bang' style of development had been achieved. After farmland for soybean cultivation matured, corn and other crops soon followed.

After the breakthroughs achieved by PRODECER during the 1970s for modern upland farming in the Cerrado region, entrepreneurial farmers from southern Brazil migrated to the Cerrado in droves. Weather characteristics (sunshine, temperature, rainfall, etc.), soil, vegetation, and socio-economic conditions in the Cerrado are diverse. Agricultural producers designed and adjusted their cultivation systems so that living matter would adapt to and disseminate over the vast and diverse environment of the Cerrado, while rapidly moving northward. Today, MT is Brazil's largest soybean-producing state, and agricultural development has spread across the horn-shaped Cerrado region in the low latitudes of northeastern Brazil. The drastic increase in the land area being used for the cultivation of the pioneer crop, soybean, illustrates this point. As already noted, during the 2000s, the soybean cultivation front reached RR, which is situated in the northern hemisphere beyond the equatorial Amazon.

The Brazilian Ministry of Agriculture, Livestock and Supply (MAPA) has taken notice of the region called MATOPIBA (shown in the dotted ellipse in Figure 3.1), which has demonstrated the highest growth rate among agricultural development areas in the Cerrado in the past decade, and which is thus also regarded as the most promising area for future development. The acronym MATOPIBA comprises the first syllables of the names

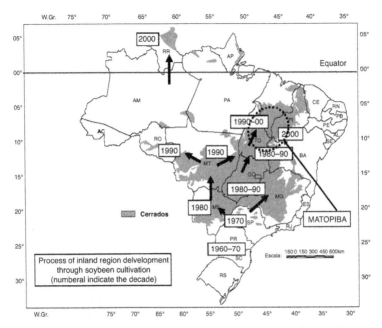

*Figure 3.1*   Progress of inland region development through soybean cultivation
*Source:* Authors based on Santo (2001).

of four states in the Cerrado region: Maranhão (MA), Tocantins (TO), Piauí (PI), and Bahia (BA). This region has received increased attention because of its established agricultural technologies, low land prices, and infrastructure: the Carajás and North–South Railroads, which can be used for transporting products to the port of São Luís; the Araguaia–Tocantins waterway (already in the federal government's strategic plan), which would open a competitive grain transportation route through Belém. Development of the Cerrado region has progressed on an especially large-scale in PI, which was once regarded as one of Brazil's poorest states.

In addition to crop production, the logistics infrastructure was developed for the transportation of bulky grains. A paved road network was gradually developed throughout the Cerrado area, and the population began to shift. In April 1997, Brazil achieved a breakthrough by developing a soybean transportation route using the Amazon River. While the United States exports grain via the Mississippi, the largest river system in North America, which flows from north to south, the Brazilians started to use the Amazon, the largest river in the world, which flows across the South American continent from west to east.

The soaring production of grain and other agricultural products led to the self-organization of strategic production bases and the formation of an upstream-to-downstream value chain, and further led to the flourishing of large-scale agri-businesses. In the 1980s, major transnational grain trade companies (GTCs) started to make inroads into the Cerrado region, which had been transformed into a bastion of agricultural production. These GTCs provided abundant funds to agricultural producers, while keeping their eye on the goal (in soybean cultivation) of vertically integrating agri-business processes from upstream (fertilizers and other inputs for production) to downstream (logistics, marketing, and export).

Meanwhile, on the other side of the globe, China had transitioned from being an exporter of soybeans to being an importer. When the country joined the WTO in 2001, it attempted a complete turnaround in its agricultural policies, and a henceforth gluttonous China was born. The international trade structure for soybeans greatly changed; today, soybeans flow from all over the world into China, which accounts for about 60 percent (about 60 million mt) of the total amount of imported soybeans worldwide. In 2009, China also turned into a net importing country for corn. Because of global expansion in the production and supply of agricultural and livestock products, including grains cultivated by Cerrado agriculture, the ever-increasing global demand for food (mainly in China) has been met. Brazil has emerged as one of the leading exporting countries, with the greatest net amount of exported agricultural products in the world.

## 3.2   Rapid growth in agricultural production and the pathway toward exports

The growth in Brazilian grain production since the beginning of the development period of Cerrado agriculture has been remarkable. Brazil's annual grain production[1] in the latter half of the 1970s, when Cerrado agriculture began, was around 40–50 million mt; this figure had risen to some 70 million mt per year by the mid-1990s. Still, the world did not focus on Brazil as a major grain-producing country. From the time that the development period began, production accelerated as technology became disseminated, and as the development of logistics networks also progressed. Ultimately, in the 2000 agricultural year, grain production exceeded 100 million mt; in 2002, the United States Department of Agriculture (USDA) pointed out that the potential output of Brazilian agriculture had been underestimated (discussed below). Brazil continued

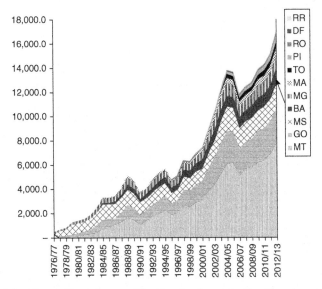

*Figure 3.2* Change in soybean cultivation land area in the Cerrado region, by state (unit: 1,000 ha)

*Notes:* (1) The Brazilian government has not provided agricultural production statistics focused on the Cerrado region: these statistics have been announced annually for the respective states. For this reason, in order to identify the trends, we assume in this publication that the total of the production figures for the ten states shown in (3) below and the Brazilian Federal District, a wide area of each of which is in the Cerrado region, represents the production (land area and production volume) for the Cerrado region. This is because grains are produced mainly on flat land or gentle slopes where mechanized farming is possible, and farming in these states and the Brazilian Federal District has been conducted mainly in the Cerrado region, which is suitable for mechanized farming. Statistics provided by the National Food Supply Company (CONAB, Companhia Nacional de Abastecimento) have been organized for each state since the 1976/1977 agricultural year, when Cerrado agriculture began, and thus are helpful. This is one of the reasons why the authors frequently make use of them. A drawback is that the statistics for the states under consideration include grain production from areas other than the Cerrado, so figures slightly larger than the actual data are being counted.

(2) The authors estimate that the margin of error (the difference between the production in the ten states plus the Brazilian Federal District and the actual production in the Cerrado area) is about 10%–20% for the major grains. The Cerrado Agricultural Research Center (CPAC) has plotted the grain production data for each county (applicable to 1,388 counties in the Cerrado region) prepared by the Brazilian Institute of Geography and Statistics (IBGE, Instituto Brasileiro de Geografia e Estatística) onto a distribution map of the Cerrado area, and aggregated the Cerrado grain production amount from 1990 to 2009 (unpublished). Comparing the data for that period with the production in (1) above, the differences are between 1% and 27% for soybean and 8.7% to 29.4% for corn, differing greatly according to the year.

(3) States counted for production in the Cerrado region: MT, GO, MS, MG, BA, MA, PI, TO, RR, RO, and the Brazilian Federal District (DF).

(4) BA, MA, PI, and TO, or the MATOPIBA region, shown within the dotted lines in Figure 3.1, constitute the Cerrado region where the development process has received the most attention, and where the land area under soybean cultivation is predicted to boom in the coming years.

*Source:* Authors, based on statistics from CONAB (National Food Supply Agency of Brazil, affiliated with the Ministry of Agriculture) (September 2011).

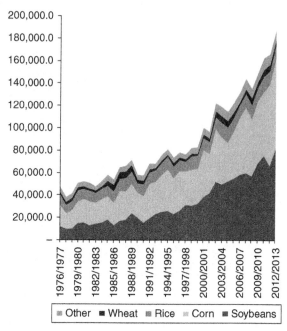

*Figure 3.3*    Growth in grain production in Brazil (unit: 1,000 mt)
Source: Authors, based on CONAB statistics (2011).

to expand production at a breakneck speed, reaching 183 million mt in grain production in 2011 (Figure 3.3).

When examining Brazil's grain production, the so-called four major grains – soybean, corn, rice, and wheat – account for more than 90 percent of its overall yield. The proportion of soybean and corn to the whole is strikingly large, with 87 percent of the total grain production accounted for by these two crops alone (fiscal year 2012/2013). As Figure 3.3 clearly demonstrates, soybean brought about the increase in Brazil's grain production. It drove agricultural development not only in the Cerrado, but also in Brazilian agriculture as a whole.

Figure 3.4 shows the change in production of the two major crops, soybean and corn, in all of Brazil and in the Cerrado region. The increase in soybean and corn production in the Cerrado region clearly raised Brazil's overall production.

In the following pages, the likely future trends in Brazil's production and export of the four major grains will be discussed, as these trends will impact on the international trade in agricultural products. These four

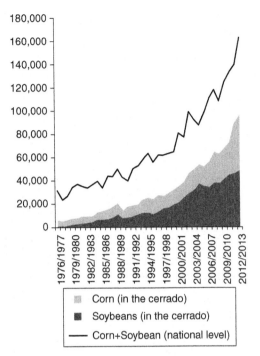

*Figure 3.4* Growth in production of soybeans and corn in Cerrado region (unit: 1,000 mt)

*Source:* Authors, based on CONAB statistics (2011). For CONAB statistics, see the notes to Figure 3.2.

grains compete with each other for production, and their demand is not mutually exclusive – such that their respective numbers can impact on one another. In order to make clear the current status of Cerrado agriculture, and to estimate its future impact on the global market, it is necessary to understand some specific features of these four crops.

## (1) Soybean

The rapid increase of soybean production in Brazil led to a surge in exports. Figure 3.5 shows the change in the volume of soybean exports from the US, the world's largest soybean exporter, and Brazil. Before Cerrado agriculture began, there was no Brazilian soybean surplus for export. The volume of soybeans exported from Brazil started to grow in the latter half of the 1990s, along with its production. In 2005–2006,

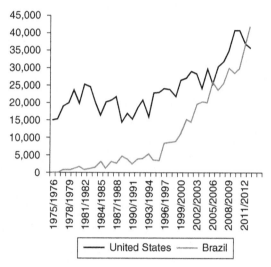

*Figure 3.5*   Growth in exported volume of soybeans by Brazil and US (unit: 1,000 mt)

*Source:* Authors, based on USDA/FAS.

Brazil was temporarily the world leader in soybean exports, although by a very small margin. Brazil again exported the world's largest volume of soybeans in 2011–2012, thus establishing itself as a soybean-exporting country comparable to the US.

Soybean oil and soymeal (soybean meal or bean cake), which is left over after the oil is squeezed out, have also been exported, and these soybean-related products, together with soybean grain, have come to account for the largest proportion of items exported by Brazilian agri-businesses. The growth of soybean production also caused a wide range of related activities to emerge in the country, as discussed in the next chapter.

The intense global demand for grains is a major factor behind the growth of soybean production in Brazil. China's increased demand is especially noteworthy. For many years, China, the country in which soybean cultivation originated, was an exporter of soybeans, but it became a soybean importer in 1996. When China joined the World Trade Organization (WTO) on December 11, 2001, it abandoned its soybean self-sufficiency policy and shifted toward a dependency on imports. Today, China accounts for 60 percent (about 60 million mt in 2010) of the total imported volume of soybean grain worldwide, which is more than 15 times the volume imported by Japan. Brazil exported 19 million

Table 3.1 Four major exporters of soybeans (average of five years in each period) (unit: 1,000 mt)

| | 1962–1966 | | 1972–1976 | | 1982–1986 | | 1992–1996 | | 2002–2006 | | 2011* | |
|---|---|---|---|---|---|---|---|---|---|---|---|---|
| 1st | US | 6,571 | US | 13,926 | US | 20,377 | US | 21,462 | USA | 27,684 | Brazil | 36,900 |
| 2nd | China | 564 | Brazil | 2,815 | Brazil | 2,256 | Brazil | 4,354 | Brazil | 21,937 | US | 34,700 |
| 3rd | Brazil | 167 | China | 255 | Argentina | 2,167 | Argentina | 2,323 | Argentina | 8,362 | Argentina | 8,900 |
| 4th | Canada | 81 | Argentina | 147 | China | 1,042 | Paraguay | 1,450 | Paraguay | 2,987 | Paraguay | 4,000 |
| Other | | 39 | | 352 | | 1,048 | | 1,708 | | 2,453 | | 6,392 |
| World total | | 7,411 | | 17,495 | | 26,891 | | 31,297 | | 63,423 | | 90,892 |
| World top exporter's share of total | | 88.5% | | 79.6% | | 75.8% | | 68.6% | | 43.6% | | 40.6% |
| World top four exporters' share of total | | 99.5% | | 98.0% | | 96.1% | | 94.5% | | 96.1% | | 92.9% |

Note: *Projection for 2011/2012 by USDA/FAS.

Source: Authors, based on USDA/FAS statistics.

mt of soybean grain to China in 2010, which accounted for 65.5 percent of the total volume exported by Brazil (29 million mt). Demand in China is largely responsible for driving soybean production in Brazil.

## (2) Corn

Trends in corn production and exports by Brazil have also shown significant changes in recent years. Statistics starting from 1975 show that the volume of corn production has tended to increase in Brazil. However, large amounts of corn were imported in the past in order to make up for the country's deficiency (Figure 3.6). Since 2005, Brazil has increased its export volume, helped by very favorable international market conditions. This development is also a result of China's influence.

China had an established agricultural policy of being self-sufficient in corn, rice, and wheat. It was also a corn-exporting country, exceeding 10 million mt per year, second only to the US. However, in line with rising standards of living, recent years have witnessed a growth in the demand for meat in China, and the amount of corn exported as feed has fallen dramatically, causing the country to shift completely to being a corn-importing country since 2009. The volume of corn imports in 2011 was 5 million mt, and the Ministry of Agriculture of China announced in October 2011 that this would reach 20 million mt in 2020. China, a major food importer, has followed the same trend with corn as it did with soybeans.

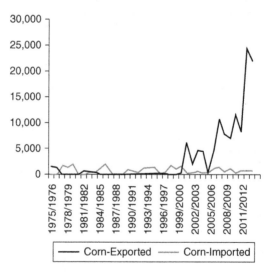

*Figure 3.6*   Brazil's external corn trade (unit: 1,000 mt)
*Source:* Authors, based on USDA/FAS.

Meanwhile, by 2010, Brazil had already risen to third in the volume of corn exported, following the US and Argentina. Brazil became the world's largest exporter of corn in 2011–2012, though the United States would retake the lead in following years. The Cerrado region is also suitable for corn cultivation, and production has doubled over the past decade (about 21 million mt in 2009). Corn is regarded as an essential crop, to be cultivated in rotation with soybeans. Corn production in the Cerrado region will continue to increase, corresponding with the development of the logistics infrastructure, and the expansion of the crop's global presence is likely to continue.

**(3) Rice**

The trade ratio (exported volume/global production) for rice has always been the lowest among the major grain crops, at 7 percent (Japan Ministry of Agriculture, Forestry and Fisheries, 2010, p. 88). Rice is basically a subsistence crop, and the overall production of rice for human consumption has been about 12 million mt per year in Brazil, which is significantly lower than the production of soybean and corn. In recent years, domestic supply and demand have been roughly in balance. The share of production in the southern rice paddy region has traditionally been high in Brazil, and the amount produced in the Cerrado region has remained at about 10 percent of national production over the past 20 years. This is because the *veranico* (dry spell during the rainy season) often affects upland rice cultivation in the Cerrado region: when it arrives at the booting stage, it may cause a severe decline in production. Rice productivity is not likely to improve in the Cerrado in the future unless irrigation is increased.

**(4) Wheat**

Wheat has traditionally been a traded agricultural product, its world exports accounting for 20 percent of its total production; this ratio is high among the major grains, following soybean (36 percent). Production areas are scattered in both the northern and southern hemispheres.

Brazil has always been a wheat importer. In recent years, nationwide production was at 5.88 million mt, while the volume of imports was 6.32 million mt (2010) – that is, nearly equal values. Brazil imports wheat from MERCOSUR (Southern Common Market) member countries, including 60 percent from neighboring Argentina and 20 percent from Uruguay. Though no government subsidy policy has been implemented for wheat, Brazil aspires to reach self-sufficiency, and new potential has now arisen in the Cerrado region.

Currently, the area of wheat cultivation in the Cerrado is about 50,000 ha, and its production accounts for only 5 percent of the national total.

However, CPAC has already accumulated 36 years of wheat research results. CPAC estimates that about 1.5 million hectares in the Cerrado region are suitable for irrigated wheat, and 3.5 million hectares for non-irrigated systems. Grain production that uses irrigation facilities is typically expensive and offers extremely low profitability. Wheat production during the rainy season (October–March) is also technically difficult due to damage from disease such as blast (*brusone*) – a fungus caused by Magnaporthe grisea in hot and humid conditions.

If wheat is planted in January, right before the end of the rainy season and after the soybean harvest, instead of during the rainy season, an average wheat yield of three mt per hectare is said to be possible during the harvesting time of the dry season. To enable this yield, it is essential to cultivate early soybean varieties (harvested in 90–100 days) that can be harvested by January. The CPAC has already selected good cultivars and is expecting to announce new ones shortly. The development of extremely early soybean varieties has the potential to transform the Cerrado into a major area of wheat production (CPAC News, 2011).

Brazil has established itself as a major grain-producing country, with soybean and corn as its key crops. In the 21st century, it has switched to being a net exporting country of grains, and today is one of the world's leading grain exporters (see Figure 3.7). Given the drastic increase in

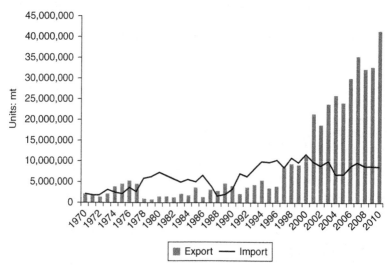

*Figure 3.7*   Evolution of Brazil's exports of major grains

*Source:* Authors, based on FAOSTAT, USDA, and OBIOVE statistics.

global demand for grains, Cerrado agriculture's potential is likely to become a reality.

One issue to overcome in this process is the development of a logistics infrastructure for transporting agricultural products.

## 3.3 Logistics infrastructure: barriers posed by vast spaces, and a new breakthrough

The vast and relatively low-priced land of the Cerrado region makes it possible to benefit from the economies of scale associated with large-scale mechanized farming. Thanks to the development of certain agricultural technologies, the region now possesses optimal conditions for grain production. However, soybean, the pioneering crop, and the corn that followed are typical commodities, in addition to their bulky volume when shipped. This made it essential to develop a transport and logistics infrastructure to reach the main export ports. Soybean cultivation has enjoyed advantages of scale; but, at the same time, the crop has suffered from difficulties related to the vastness of the land.

Paved road networks were continuously extended toward inland areas, following the extension of the agricultural frontier. But the roads faced a trio of difficulties: the paving work was unable to keep pace with the agricultural development; the paved roads could not withstand the weight of 30 mt trucks, and quickly deteriorated; and, above all, the loading ports were too far away, meaning that the transport costs were too high. For example, in the case of MT, which is the largest soybean-producing state, the distance by road from the major production zones to Brazil's largest grain-exporting port, Paranaguá, is 2,500 km – almost equivalent to the distance between the northern and southern ends of Japan. The land transport cost in 1997 was as high as US$80 per mt. Road transport costs had risen to the point that they posed a major obstacle to international competition. There was a need to link Brazil's transportation infrastructure by establishing a multimodal transportation system and developing transportation routes.

On April 12, 1997, the establishment of a soybean export route via the Amazon River represented a breakthrough for Cerrado agriculture. President Fernando Henrique Cardoso attended the completion ceremony of Itacoatiara Port, located midstream in the Amazon, and the Brazilian media were effusive in their praise of the new logistics system. The route of this new multimodal transport system, as well as other transport routes are shown in the Figure 3.8.

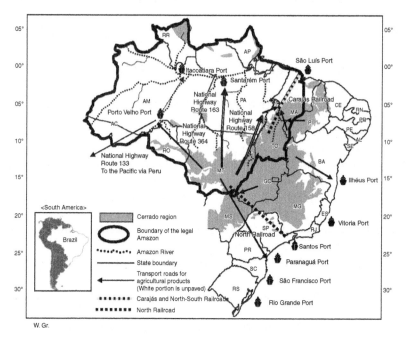

*Figure 3.8*   Major ports for exporting soybeans cultivated in the Cerrado
*Source:* Authors.

The Cerrado soybean produced on the Parecis Plateau in north-western MT is carried on 30 mt trucks for 900 kilometers northwest along National Highway 364 to Porto Velho Port in RO, in the upstream area of the Amazon River. Barge convoys are formed (with a maximum load of 18,000 mt, equivalent to 600 trucks), and the products are transported 1,050 kilometers from there to Itacoatiara Port, midstream along the Amazon. The convoys are guided by global geographical positioning system (GPS). The port is 270 km downstream from Manaus, the state capital of Amazonas (AM), where water depths vary from 22 to 28 m. For this reason, piers are fixed at two positions on land and two underwater, allowing the docking of 60,000-ton grain vessels (Panamax vessels, which can sail through the Panama Canal). These vessels sail directly from midstream along the Amazon to Europe and Asia.

The Maggi Group (now the Amaggi Group) of Brazil, led by André Maggi, an entrepreneur from Paraná, in the south, who made inroads into Parecis Plateau in northwestern MT for soybean production in 1986, developed the route. Thus, it was an outsider who conceived this novel idea. His son Blairo Maggi succeeded him and was nicknamed Brazil's Soybean King. Today, he heads one of the nation's largest soybean producers.

Comparing the costs as of February 1998, shipments to Rotterdam using conventional truck transportation via Paranaguá Port cost US$108 per mt, while those using the new route cost US$84 per mt (based on an on-site survey conducted at the time by Yutaka Hongo, one of the authors). The time required for transport was also cut from 30 days by the conventional route to 23 days for the new route.

After confirming the reduced costs of the export corridor via the Amazon, Cargill, a major grain company, built a loading port in Santarém, midstream along the Amazon, in 2003 (Figure 3.9). This port is situated at the terminal of National Highway 163, which crosses the center of the Amazon region.

Among other efforts to develop export routes for grain produced in the Cerrado, development work is under way to extend the Carajás and North–South Railroads for grain transportation, and to use rivers in this area, as well as to pave roads that connect the export terminals. The Pacific shipment route, via Peru, is being explored by the Initiative for the Integration of the Regional Infrastructure of South America (IIRSA). Soybean exports from regional ports and harbor facilities in northern Brazil have surged (Figure 3.9), accounting for 15 percent of current overall exports (Figure 3.10).

Even now, Brazil largely depends on roads (trucks) for domestic transport of cargo, at a 52 percent ratio (2011); the ratio for railroad transport is 30 percent, with 13 percent for river transport. (Minisiterio dos Transportes 2012) For this reason, in 2007 the Ministry of Transport

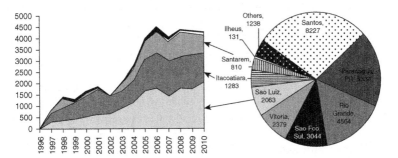

*Figure 3.9* Growth in exported soybean volume by port in the Northern region (unit: 1,000 mt)

*Source*: Prepared by the authors based on ANEC statistics (2011).

launched the National Plan of Logistics and Transportations (PNLT, Plano Nacional de Logistica e Transportes) to increase railroad transportation to 25 percent and water-based transportation to 29 percent, in which the use of the Amazon River was key (ANTAQ, 2007). The improvement of logistics will further stimulate agricultural production in the Cerrado.

On the other hand, concerns have been raised within and outside Brazil that the development of a logistics infrastructure in the Amazon region will trigger a negative chain of development, and promote the illegal logging of the Amazon's tropical forest and the inception of soybean cultivation in the area. However, there is no direct causal relationship between Cerrado agricultural development and deforestation in the Amazon. Historically, the Cerrado region has served as a buffer to the Amazon region, and has been effective in suppressing large-scale agricultural development in the Amazon. The reinforcement of surveillance systems is called for in order to prevent harm to forests in the Amazon region, as caused by logistics infrastructure development for shipping grain produced in the Cerrado.

Chapter 5 discusses the current situation surrounding soybean cultivation in the Amazon region and capabilities for suppressing illegal logging, along with technological cooperation using Japan's monitoring satellites.

## 3.4    Agri-business by transnational grain trade companies (GTCs)

As stated above, in the 1980s major GTCs, often called 'grain majors,' started making inroads into the Cerrado region, which had been transformed into a leading area for agricultural production. These GTCs supplied abundant funds to agricultural producers, always keeping their eyes on their main goal: to vertically integrate soybean agri-business processes from upstream (fertilizers, agrichemicals, and other materials for production) to downstream (logistics, marketing, and export).

This dynamic process was very visible, because gigantic grain silos bearing the names of big GTCs such as Cargill, Bunge, ADM (Archer Daniele Midlands), and Coinbra (Louis Dreyfus Group) sprang up across the Cerrado region in towns where harvested soybeans were collected. These four grain majors now have an enormous presence in Brazil.

GTCs have been active in Brazil for many years. Bunge, from the Netherlands, and Louis Dreyfus, from France, made inroads into Brazil in the early 1900s, and the US giant Cargill did likewise in the 1960s. In the early stages, these enterprises handled a wide range of agricultural products, and held a status equal to that of many of the domestic food companies in Brazil and other companies that had entered the country; it was a long time before the four companies made their presence felt as grain majors. A big change occurred along with the development of Cerrado agriculture. During the 1980s these GTCs became noticeably active in Brazil, and their activities further intensified during the 1990s.

Grain production volume doubled in Brazil during the 1990s, from 58 million mt at the beginning of the decade to 122 million mt in 2002. The increase in soybean production was especially rapid during this period, from 15 million to 52 million mt; soybeans became the most important crop in Brazil, accounting for 25–42 percent of total grain production. In the 1990s, the GTCs had either merged with or acquired domestic companies and others in the soybean industry, and expanded their direct investments to strengthen their presence. Noteworthy examples include Dreyfus's 1996 acquisition of Anderson Clayton; Bunge's 1997 acquisition of Ceval, the largest soybean oil extracting company in Brazil; and the acquisition of the soybean divisions of Sadia in 1997 by ADM (a US company that arrived in Brazil later than the others). The acquisition of domestic soybean oil extracting companies by the GTCs and their development of storage facilities and export base networks largely reorganized the industry. Today's four powerful GTCs in Brazil – Bunge, Cargill, ADM, and Louis Dreyfus (Coinbra), the so-called "ABCD" – were established as a result of these moves.

A number of background factors helped to intensify the GTCs' investment activities during the 1990s. These background factors included the stabilization of the Brazilian macro-economy, which was achieved while implementing strategies by the GTCs for international market enclosure. In addition, the advantages of the Brazilian production locations became noticeable – advantages such as the vast land area and abundant resources, improved production technologies, and the difference in the harvesting period from the northern hemisphere.

The GTCs also supplied abundant overseas funds to farmers as pre-production loans ('green soybean,' which is *soja verde* in Portuguese), and distributed these farmers' products to international markets. In this way, the GTCs established contract cultivation systems with farmers. These strategies were also in line with the policies for attracting foreign capital and expanding the exports promoted by the Brazilian government, and were effective in supplementing deficiencies in agricultural credit funding.

At the same time that the GTCs were expanding their investments, the volume of soybean grain exported by Brazil was surging: from 4 million mt in the mid-1990s to 21 million mt in 2002/2003, a volume comparable to that exported by the US. Of the volume of soybean grain exported from Brazil in 1998, the portion handled by the GTCs accounted for 37 percent. Their share rose to 50 percent in 2002 (74 percent when exports by affiliated companies are included; see Figure 3.10), and their market share reached 60 percent in 2005. The GTCs promoted vertical integration of businesses (food chain development) from upstream-to-downstream processes, covering fertilizer production, oil extraction, food production, logistics, marketing, and export.

The GTCs have become more involved in corn in recent years, having registered the increase in Brazilian corn production and the country's switch to being a corn exporter.

Apart from the GTCs of the West, the activities of the domestic Brazilian Amaggi Group are worthy of note. This group of companies has practiced soybean production, oil extraction, and export businesses in Mid-Western and Northern Brazil. It owns the Itacoatiara Port and annually exports 10 million mt of soybean grain; furthermore, it has established close ties with Japanese trading companies.

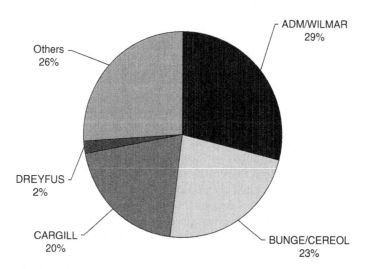

*Figure 3.10*   Shares of soybean grain exported by grain majors (2002)

*Source:* Authors, based on Transcar Group 2002/2003 statistics and interviews with Japanese trading companies.

The activities of the GTCs have had a wide range of influences on the development of Cerrado agriculture, such as: (1) developing agri-businesses that participate in the global market; (2) introducing foreign investment; (3) acquiring foreign currency through exports; and (4) supplying production funds to supplement deficits in financing systems (Toyoda, 2011). The agri-business food chain has been transformed into an oligopoly; however, the early years of the development period of Cerrado agriculture coincided with Brazil's period of economic difficulty. By lending funds for production, these GTCs supported farmers who had defaulted and had been shut out of loan systems.

Given the enormous agricultural potential of the Cerrado, as discussed below, it is reasonable to expect that not only the traditional GTCs but also Amaggi and other domestic Brazilian companies, as well as Asian companies, will participate in the logistics and export arena. Through diversification and the widening of alternatives of crops for agricultural producers, Cerrado agriculture will continue to be developed.

## 3.5   Cerrado agriculture in the process of diversification

As Cerrado agriculture developed, there was diversification in the agricultural products being cultivated. In addition to soybeans, corn, rice, common beans (*Phaseolus spp.*, popularly called 'feijao'), wheat and other grains, coffee, cotton, rubber trees and vegetables (carrots, potatoes, industrial-use tomatoes, and others) are now being cultivated. Recently, the production of sugarcane for ethanol has also expanded. Irrigation agriculture has been introduced in some places.

Since its early years, Cerrado agriculture had been conducted on vast, non-irrigated fields. More recently, irrigation achieved by large-scale center pivot-type irrigation facilities has expanded into areas with access to water. In addition to farming, beef cattle (usually Zebu cattle) have been pastured on vast tracts of improved pasture land. Some regions have undergone large-scale planting of eucalyptus trees. Cerrado agriculture has become extensively diversified, to cover grain, cotton, livestock rearing, fruit, vegetables, bio-energy crops, and silviculture. The production of major agricultural products other than grain in the Cerrado region is discussed below, looking back on the 40-year development process of agriculture and livestock rearing.

### (1) Improved pasture

At any one time, there are 205 million head of commercial cattle being raised in Brazil, which is the largest figure in the world (FAO,

2010). Since 2004, Brazil has also boasted the world's largest volume of exported beef: its market share has reached about 20 percent. Since the National Beef Cattle Research Center (EMBRAPA Beef Cattle) was established in 1975 at Campo Grande, the state capital of MS in the Cerrado region, productivity has increased due to the improvement of beef cattle breed and pasture grass, and innovations in breeding and administration technologies. Beef cattle production in the Cerrado region is mainly accomplished by pasturing, rather than keeping cattle in barns, as in Japan, or in feedlots (fattening grounds where they are given blended artificial feed), as is generally practiced in the United States. For this reason, meat produced in Brazil has been referred to as *carne verde* (green meat).

Broadly established tropical grass pasture areas are spread over the Cerrado region, making it possible to benefit from economies of scale. EMBRAPA Beef Cattle estimates the pasture area in the Cerrado region to be 55 million ha (almost equivalent to the 58 million ha in the Mid-Western region of Brazil, as estimated by IBGE). According to IBGE, 75 percent of the pasture land in the Mid-Western region (about 45 million hectares) has already been 'improved.' This accounts for 45 percent of the overall improved pasture area in Brazil. What had been only 9 million ha of improved pasture land in the Mid-Western region in 1970, prior to the start of Cerrado agriculture, had increased five-fold as of 2006.

Improved pasture refers to pasture land where the soil has been improved and high-quality grazing grass cultivars have been planted, instead of the traditional form of grazing in the native Cerrado. The Cerrado soil lacks organic substances, so cattle dung is used to improve it. Pasture land can be easily converted to crop fields. In order to ensure sustainable agricultural development in the Cerrado region and to improve pasture and fields for grain production, researchers even recommend introducing systems for rotational land use. In recent years, EMBRAPA has supported a significant amount of research on the integration of crop, livestock, and forestry agricultural systems (*integração lavoura, pecuária e floresta* (ILPF)), in order to optimize land use.

Although improved pasture is not counted as grain cultivation land, it could result in farmland. The USDA has indicated that pasture land can easily be converted to farmland; thus, the area of cultivated land in Brazil could surpass that of the US. This point is discussed in the final section of this chapter.

## (2) Coffee

Brazil has always been associated with coffee, which is a historically important crop. The coffee production that emerged along the coast of Rio de Janeiro (RJ) flourished in the late 19th to early 20th centuries in São Paulo (SP). Since then, it has been the leading export product for the Brazilian economy. Today, coffee remains a major agricultural export for the country, but its relative importance has been dramatically reduced, and the main production ground has also significantly shifted.

The land area used for coffee cultivation in Brazil has remained almost unchanged over the past decade, at about 2.1 million ha (CONAB, 2011), but the cultivation grounds have shifted from the traditional areas in SP and PR. Today, land areas used for coffee cultivation in the latter two states account for only 13 percent of the national total (280,000 ha). On the other hand, the cultivation area has increased in the Cerrado region, particularly in MG and BA.

This increase is particularly conspicuous in MG, which now accounts for 50 percent of the national total (1.05 million ha). Coffee production in Brazil has shifted to the Cerrado area; moreover, the Cerrado region has been transformed into a major coffee production ground.

## (3) Silviculture

Another form of large-scale agriculture is forestry (silviculture). Nationwide, the total land area planted with trees is 6.5 million ha (ABRAF, 2011). In Brazil, a significant number of eucalyptus trees brought in from Australia have been planted, accounting for 73 percent of the total tree-planted area in Brazil. Looking at the distribution of areas with planted trees, MG stands out, with 24 percent of the national total, followed by SP (19 percent) and PR (13 percent). The land area being used for silviculture in the Cerrado states totals 3 million ha, or 46 percent of the total land area planted with trees.

## (4) Sugarcane

Sugarcane produced in Brazil has been used for producing sugar and ethanol for automotive fuel. The land area under sugarcane cultivation has rapidly increased, thanks to strong domestic and overseas demand. The amounts of sugarcane produced for manufacturing sugar and ethanol were almost equal as of 2011, although the exact ratio fluctuates due to international market trends. The overall land area under sugarcane cultivation in Brazil is 8.43 million ha (CONAB, 2011), and SP traditionally stood out, with 4.43 million ha (53 percent of the total) in 2011. In recent years, however, the production ground has rapidly spread to neighboring states, mainly to MG (760,000 ha), MS (480,000 ha), and GO (670,000 ha) (Figure 3.11). The sugarcane cultivation front is expected to continue to spread into the Cerrado region, supported by the strong demand for ethanol as an automotive fuel.

This review of the changes in land use in the Cerrado region makes the recent trends clearer. The majority of farmland in the Cerrado region is used as pasture ground, and the ratio of improved pasture has rapidly increased. The land area under cultivation for annual crops (grain, vegetables, and so on) accounted for only about 10 percent of the total in 2002 (see Chapter 8). Much of the subsequent increase in annual crop production was attained by improvements in yield (see Chapter 6), since the area of annual crop cultivation increased only slightly between 2002 and 2006 in the 11 states located in the Cerrado.[2]

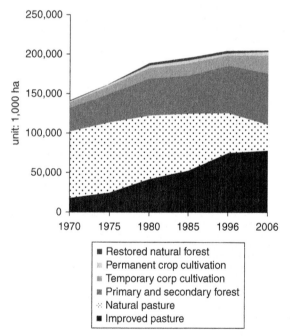

*Figure 3.11* Changes in areas of land use in the 11 states comprising the Cerrado

*Source:* Authors, based on IBGE statistics.

*Table 3.2* Land use trends in the 11 states comprising the Cerrado* (unit: 1,000 ha)

|  | 1970 | 1975 | 1980 | 1985 | 1996 | 2006 |
|---|---|---|---|---|---|---|
| Improved pasture | 17,500 | 24,989 | 41,666 | 52,806 | 75,186 | 78,040 |
| Natural pasture | 84,741 | 88,558 | 80,463 | 72,362 | 50,668 | 33,358 |
| Primary and secondary forest | 28,838 | 36,033 | 46,810 | 47,716 | 59,124 | 64,560 |
| Annual crop cultivation | 7,979 | 10,888 | 14,078 | 16,318 | 13,595 | 22,558 |
| Perennial crop cultivation | 1,779 | 2,069 | 3,313 | 3,821 | 3,375 | 5,080 |
| Restored natural forest | 369 | 835 | 2,295 | 2,677 | 2,420 | 1,663 |
| Total area (11 states) | 141,205 | 163,372 | 188,626 | 195,700 | 204,367 | 205,259 |

*Note:* *See Note (3) to Figure 3.2.

*Source:* Authors, based on IBGE statistics.

Although the land areas used for coffee cultivation and silviculture have grown, their overall ratios are extremely small, at 2 to 3 percent.

It is not hard to imagine the improved pasture land in the Cerrado region being converted into cultivation grounds for grain or sugarcane, in view of the growing global demand for grain and for sugarcane as a source material for biological fuel (ethanol).

In short, the Cerrado region holds enormous potential for grain production, biological fuel crops, silviculture, and the production of beef cattle, which will continue to offer economies of scale if logistics infrastructure development continues to progress (Table 3.2).

## 3.6 Brazil as a major country based on agriculture and the potential of Cerrado agriculture

In December 2001, the USDA presented an analysis of Brazil's competitiveness with regard to exports, in a report that compared the grain production competitiveness of the US with Latin American countries (USDA, 2001). Subsequently, the USDA pointed out that Brazil's agricultural potential had been underestimated (USDA, 2003). Both reports attracted considerable attention in Brazil. The latter report stated that

Brazil's potential farmland was conservatively estimated at 170 million ha (95 million ha in the Cerrado region, and 75 million ha in the Amazon and other areas, including farmland converted from pasture land). If this potential land were to be developed into farmland, the total cultivated land area in Brazil would reach 210 million ha, which far surpasses the 174 million ha of cultivated land in the US. Regarding soybeans, the land area under cultivation (24 million ha as of 2011) could be expanded to 50–100 million ha in the coming decade. The USDA (2003) noted various favorable conditions, such as excellent agricultural management, agricultural technologies at an international standard, low-priced and abundant land, and investment by the government and private sector, as well as GTCs' engagement in the entire process. The USDA and American universities later published papers pointing out potential bottlenecks in the expansion of soybean production in Brazil, and papers that discussed the strong and weak points of Brazil and the US in terms of soybean production.

Meanwhile, Brazil already has the world's largest trade surplus of agricultural products,[3] and in this regard has become the world's most powerful country on the basis of agricultural production. According to the *Statistical Yearbook* published by the United Nations Food and Agriculture Organization (FAO), since 2002 Brazil has remained at the top of the world rankings for the net amount of exported agricultural products (the amount obtained by subtracting the total amount of imports from the total amount of exports). According to the most recent values, shown in the 2011 version of the Yearbook (actual results for 2008), Brazil's trade surplus ranked first among 183 nations at US$47.4 billion, far above the second-ranked US's US$35.8 billion.[4]

Figure 3.12 shows the breakdown of of Brazil's agricultural product exports' value in 2000 and 2010. Soybean-related products remained at the top throughout, but notable changes have been seen in recent years. The export ratios of sugar/ethanol and livestock products (beef, chicken, and pork) have risen, while those of forestry products (lumber and pulp) have tended to decline. Brazil's agricultural product exports have shifted somewhat, showing a higher emphasis on soybean and meat, while the ratios of its traditional leading agricultural products, such as coffee, cacao, bananas, and other tropical crops, have fallen drastically.

Having realized large-scale modern upland farming in the tropical Cerrado, and having dramatically boosted levels of food safety and security, Brazil has come to be regarded by African and Latin American countries as a model for agricultural development.

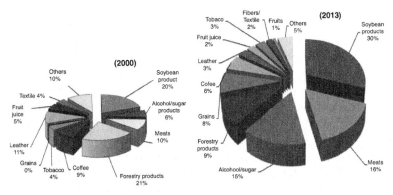

*Figure 3.12* Breakdown of Brazil's major agri-business exports (value)

*Source:* Authors, based on Ministry of Agriculture (2011 and 2014) chart, 'Major categories of 450 items exported by Brazil.'

## Notes

1. The grain statistics published by the National Food Supply Company (CONAB), which is affiliated with the Ministry of Agriculture of Brazil, cover the production of 15 agricultural products: soybeans, corn, rice, wheat, feijão beans, buckwheat, peanuts, sunflowers, castor-oil plants, sorghum, rye, common wild oats, triticale, cotton, and rapeseed.
2. As regards the 11 states located in the Cerrado, see notes to Figure 3.2.
3. Here, the strength of a country's agriculture is indicated on the basis of the net amount of agricultural products traded, instead of the exported volume converted to energy. Specifically, the amount was calculated by using the following equation: (Amount of agricultural products exported) – (Amount of agricultural products imported) = Net amount of agricultural products exported (surplus).
4. Incidentally, Japan ranks at the bottom (183rd), with a trade deficit of US$53.9 billion, even larger than China's deficit of US$45 billion.

## Bibliography

ABRAF (Brazilian Association of Forest Plantation Producers) (2011) *ABRAF Statistical Yearbook 2011: Base Year 2010*, http://www.cnr.ncsu.edu/sofac/ABRAF_Statistical_Annual_Report_2011_English.pdf, date accessed May 29, 2014.

Alencar, Sávio Borges, Alisson Diego do Nascimento Neri and Eliane Pinheiro de Souza (2011) Encadeamentos do setor agropecuário brasileiro no período de 1997 a 2007. *Revista Política Agrícola*, 20(4).: pp.58–68

Almeida, Maria Geralda (ed.) (2005) *Tantos cerrados* (Texas: Editora Vieria).

ANEC (Associação Nacional dos Exportadores de Cereasis) (2011) *ANEC Statistics: Evolucao das exportacoes por Porto 2011.*

ANTAQ (Agencia Nacional de Transportes Aquaviários) (2007) *Anuario estatistico 2007.*

Aquino, Fabiana de Gois and Maria Cristina de Oliveira (2006) 'Reserva legal no bioma Cerrado: uso e preservação,' EMBRAPA Cerrados, Document No. 158, http://www.infoteca.cnptia.embrapa.br/bitstream/doc/570268/1/doc158.pdf, date accessed May 29, 2014.

Barreto, Clarissa de Araújo (2007) 'Agriculture and environment: perceptions and practices of soybean-planters in Rio Verde-GO,' Masters Dissertation, Universidade de São Paulo, http://www.teses.usp.br/teses/disponiveis/90/90131/tde-14082007-231915/pt-br.php, date accessed May 29, 2014.

Barros, Jose Roberto Mendonça (ed.) (2013) 'Soja intacta: Uma visão econômica dos benefícios da adoção da nova tecnologia,' MBAgro, http://monsanto.com.br/sala_imprensa/estudos/pdf/estudo-soja-intacta-mbagro.pdf, date accessed May 29, 2014.

Bernardes, Júlia Adão (2007) 'Agricultura moderna e novos espaços urbanos no Cerrado Brasileiro,' *Revista Tamoios*, 3(1), http://www.e-publicacoes.uerj.br/index.php/tamoios/article/view/618, date accessed May 29, 2014.

Brandão, Antonio Salazar Pessoa, Gervásio Castro de Rezende and Roberta Wanderley da Costa Marques (2005) 'Agricultural growth in Brazil in the period 1999–2004: Outburst of soybeans and livestock and its impact on the environment,' Instituto de Pesquisa Econômica Aplicada (IPEA), Discussion Paper No. 1103, http://econpapers.repec.org/paper/ipeipetds/1103.htm, date accessed May 29, 2014.

Brasil, Daniel, F.M.P. Teixeira and M.S. Pimentel (2006) *A colonização do cerrado: savanas e celeiro do mundo* (Sao Paulo: Prêmio).

Carranco, Adolfo A. Laborde (2011) 'Official development assistance of Japan in Brazil: the link between aid and diplomatic ties'. *Latin American Policy*, 2(1), pp. 3–12.

Centro De Sensoriamento Remoto (CSR/IBAMA) (2009) 'Relatório técnico de monitoramento do desmatamento do bioma Cerrado, de 2002 a 2008: Dados revisados,' Instituto Brasileiro Do Meio Ambiente E Dos Recursos Naturais Renováveis (IBAMA), (Brasilia: Ministério do Meio Ambiente).

Companhia Nacional de Abastecimiento (CONAB) (2011) *Pesquisa de Safras e Informacoes Geographicas da Agricultura Brasileira* (Brasilia: CONAB) http://www.conab.gov.br/conteudos.php?a=1534&t=2, date acessed July 8, 2015

CPAC News; July 21, 2011 http://www.cpac.embrapa.br/noticias/noticia_completa/333/, date accessed July 8, 2015. (Brasilia: CPAC)

Damiani, Octavio (2003) 'Effects on employment, wages, and labor standards of non-traditional export crops in northeast Brazil.' *Latin American Research Review*, 38(1), pp. 83–112.

Diniz, Bernardo Palhares Campolina (2006) 'The great Brazilian woodland-savanna: geopolitics and economics,' Doctoral Thesis, Universidade de São Paulo, http://www.teses.usp.br/teses/disponiveis/8/8136/tde-18062007-152913/pt-br.php, date accessed May 29, 2014.

Faleiro, Fábio Gelape and Austeclinio Lopes de Farias Neto (2009) *Savanas: demandas para pesquisa* (Planaltina: EMBRAPA Cerrados).

Faleiro, Fábio Gelape and Austeclinio Lopes de Farias Neto (ed.) (2009) *IX simpósio nacional sobre cerrado e II simpósio internacional sobre savanas tropicais: Menções honrosas* (Planaltina: EMBRAPA Cerrados).

Faleiro, G.F. and E.S. Sousa (eds) (2007) *Pesquisa, desenvolvimento e inovação para o Cerrado* (Planaltina: EMBRAPA Cerrados).

FAO (Food and Agriculture Organization of the United Nations) (2010) *FAO Statistical Yearbook 2010*, http://www.fao.org/economic/ess/ess-publications/ess-yearbook/ess-yearbook2010/en/, date accessed May 30, 2014.

FAO (2013) The state of food insecurity in the world: The multiple dimensions of food security, http://www.fao.org/publications/sofi/en/, date accessed May 30, 2014.

Federação das Indústrias do Rio de Janeiro (2011) *Índice FIRJAN de Desenvolvimento Municipal – Ano Base 2009*.

Figueiredo, Margarida Garcia de, Alexandre Lahós Mendonça de Barros and Joaquim José Martins Guilhoto (2005) Relação econômica dos setores agrícolas do Estado do Mato Grosso com os demais setores pertencentes tanto ao Estado quanto ao restante do Brasil. *Revista de Economia e Sociologia Rural* [online], 43(3), pp. 557–575.

Filho, Naercio Menezes and Luiz Scorzafave (2009) 'Employment and inequality of outcomes in Brazil,' Insper Working Paper, Insper Instituto de Ensino e Pesquisa.

Flaskerud, George (2003) 'Brazil's soybean production and [its] impact,' Northern Dakota State University Extension Service, http://www.ag.ndsu.edu/pubs/plantsci/rowcrops/eb79.pdf, date accessed May 30, 2014.

Fujita, Yasuo (2011) 'What makes the Bangladesh Local Government Engineering Department (LGED) so effective? complementarity between LGED capacity and donor capacity development support,' JICA-RI Working Paper No. 27.

Gates, Bill (2011) 'Innovation with impact: Financing 21st-century development,' November 3, http://www.gatesnotes.com/Development/G20-Report-Innovation-with-Impact, date accessed May 30, 2014.

Guimarães, Fábio de Macedo Soares (1949) *O Planalto Central e o problema da mudança da capital do Brasil: e, Trabalhos de campo e de gabinete da segunda expedição geográfica ao Planalto Centra* (Instituto Brasileiro de Geografia e Estatística, Conselho Nacional de Geografia).

Hosono, Akio, Shunichiro Honda, Mine Sato, and Mai Ono (2011) 'Inside the black box of capacity development' in H. Kharas, K. Makino and W. Jung (eds), *Catalyzing development: A new vision for aid* (Washington, D.C.:Brookings Institution Press).

Inocêncio, Maria Erlan and Manoel Calaça (2010) Estado e território no Brasil: reflexões a partir da agricultura no Cerrado. *Revista IDeAS*, 4(2), pp. 271–306.

Informa Economics FNP (2011) *Anualpec 2011: Anuário da pecuária Brasileira*.

Instituto Brasileiro de Geografia e Estatística (1979) *Região do cerrado: uma caracterização do desenvolvimento do espaço rural* (Rio de Janeiro: IBGE).

Instituto Brasileiro de Geografia e Estatística (2010) 'Indicadores de Desenvolvimento Sustentável Brasil 2010.' *Informação Geográfica*, 7 (Rio de Janeiro: Estudos e Pesquisas).

Instituto Brasileiro de Geografia e Estatística (2010) *Atlas nacional do Brasil: Milton Santos*.

Instituto Brasileiro de Geografia e Estatística (2012) *Atlas do espaço rural Brasileiro*.

Inter-American Development Bank (IDB) (2010) *The age of productivity: Transforming economies from the bottom up* (Washington, D.C.: IDB).

JICA (2002) *Japan–Brazil Agricultural Development Cooperation programs in the Cerrado region of Brazil: Joint evaluation study: General report* (Tokyo: JICA).

Kharas, Homi, Koji Makino, and Woojin Jung (eds) (2011) *Catalyzing development: A new vision for aid* (Washington, D.C.:Brookings Institution Press).

Klink, Carlos A. and Ricardo B. Machado (2005) 'A conservação do Cerrado Brasileiro.' *Megadiversidade*, 1(1), pp. 147–156.

Leal, Maria Luísa Campos Machado (1985) 'Transformações no cerrado e relações sociais de produção.' *Revista da Fundação João Pinheiro*,, 15(5/6), pp. 39–51.

Marchetti, Delmar and Antonio Dantas Machado (eds) (1979) *Cerrado: uso e manejo: V Simpósio sobre o Cerrado: EMBRAPA-CPAC, CNPq.* (Brasilia Editerra).

Margulis, Sergio (2004) 'Causes of deforestation of the Brazilian Amazon,' World Bank, Working Paper No. 22.

Martinelli, Luiz, Rosamond Naylor, Peter M. Vitousek, and Paulo Moutinho (2010) 'Agriculture in Brazil: impacts, costs, and opportunities for a sustainable future.' *Current Opinion in Environmental Sustainability*, 2(5–6), pp. 431–438.

Ministério da Agricultura, Pecuária e Abastecimento (2013) *Projeções do agronegócio: Brasil 2012/2013 a 2022/2023*, Projeções de Longo Prazo (Brasilia: Ministério da Agricultura, Pecuária e Abastecimento ).

Ministério dos Transportes (2007) *Plano Nacional de Logística e Transportes: Relatório Executivo.* (Brasilia: Ministerio dos Transportes)

Ministério dos Transportes (2012) *Projeto de Reavaliação do Estimativos e Metas do PNLT.*(Brasilia: Ministerio dos Transportes)

http://transportes.gov.br/images/2014/11/PNLT/2011.pdf, date accessed July 8, 2015.

Ministério dos Transportes (2013) *Transportes 2013.* (Brasilia: Ministerio dos Transportes)

Ministry of Agriculture, Forestry and Fisheries (Japan) (2010) *Annual report on food, agriculture, and rural areas in Japan.*

Mizobe, Tetsuo (2011) 'Formation of agricultural clusters in [the] Cerrado and their development effects using Luis Eduardo Magalhaes in Bahia as an example,' Background Paper for Project History on the Cerrado Agricultural Development Project in Brazil.

Muller, Charles Curt (2003) 'Expansion and modernization of agriculture in the Cerrado: the case of soybeans in Brazil's Center-West.' *Série Texto para Discussão*, 306.

OECD (2010) *Tackling inequalities in Brazil, China, India and South Africa: The role of labour market and social policies* (OECD Publishing).

Oliveira, Luiz Antonio Pinto de and Antônio Tadeu Ribeiro de Oliveira (eds) (2011) Reflexões sobre os deslocamentos populacionais no Brasil. *Instituto Brasileiro de Geografia e Estatística, Informação Demográfica e Socioeconômica*, 1 (Rio de Janeiro: Estudos & Análises).

Oliveira, Paulo S. and Robert J. Marquis (2002) *The Cerrados of Brazil: Ecology and natural history of a neotropical savanna* (New York: Columbia University Press).

Ortega, Antonio César and Clesio Marcelino Jesus (2011) 'Território café do cerrado: transformações na estrutura produtiva e seus impactos sobre o pessoal ocupado.' *Revista de Economia e Sociologia Rural* [online], 49(3), pp. 771–800.

Pereira, Roberto Carvalho and Luiz Carlos Bhering Nasser (1996) *Anais do VIII Simpósio sobre o Cerrado: Biodiversidade e producao sustentavel de alimentos e fibras nos Cerrados* (Planaltina: EMBRAPA/CPAC).

Pinto, Maria Novaes (1990) *Cerrado: caracterização, ocupação e perspectivas* (Editora Universidade de Brasília).

Plato, Gerald and William Chambers (2004) 'How does structural change in the global soybean market affect the U.S. price?,' OCS 04D-01, Electronic Outlook Report from the Economic Research Service, United States Department of Agriculture.

Rezende, Gervásio Castro de (2003) 'Agricultural growth and agrarian structure in the Brazilian Cerrado: the role of land prices, natural resources and technology.' *Brazilian Review of Economics and Agribusiness*, 1(1), pp. 117–129.

Rohter, Larry (2007) 'Scientists are making Brazil's savannah bloom.' *New York Times*, October 2.

Roessing, Antonio Carlos and Joelsio José Lazzarotto (2004) 'Criação de empregos pelo complexo agroindustrial de soja,' Documentos 233 (London: EMBRAPA).

Rydlewski, Carlos and Clarice Couto (2011) 'Por que a Embrapa é o nosso Google.' *Época Negócios*, January 31.

Samuel, Frederico (2010) *O novo tempo do cerrado: expansão dos fronts agrícola e controle do sistema de armazenamento de grão* (São Paulo: Annablume/Fapesp).

Sano, Edson Eyji, Riberto Rosa, Jorge Luís Silva Brito, and Laerte Guimarães Ferreira (2008) 'Mapeamento semidetalhado do uso da terra do Bioma Cerrado.' *Pesquisa Agropecuária Brasileira*, 43(1), pp. 153–156.

Santo, Benedito Rosa do Espírito (2001) *Os caminhos da agricultura brasileira* (São Paulo: Evoluir).

Santos, Rosselvelt José (2008) *Gaúchos e mineiros do cerrado: Metamorfoses das diferentes temporalidades* (EDUFU).

Schnepf, Randall D., Erik Dohlman, and Christine Bolling (2001) 'Agriculture in Brazil and Argentina: developments and prospects for major field crops,' USDA International Agriculture and Trade Report, Outlook No. WRS-013.

Silva, Roberto Fray, Giana de Vargas Mores, and Bruno Rógora Kawano (2013) 'Logística – sugesões para a superação de gargalos.' *Agroanalysis*, 33(1), pp. 17–18.

Solbrig, O.T. and M.D.Young (eds) (1993) *The world's savannas: Economic driving forces, ecological constraints and policy options for sustainable land use* (New York: Taylor & Francis).

Spehar, C.R., P.I.M Souza, and FAO (1999) *Sustainable cropping systems in the Brazilian Cerrados* (FAO).

The Economist (2010) 'Brazilian Agriculture: The Miracle of Cerrado.' August 26.

Toyoda, Takashi (2001) *Agribusiness no kokusaikaihatsu: Nosanbutsu boeki to takokuseki kigyo* [International development of agribusiness: trade of agro-food products and multinational corporations]. (Tokyo: Noson Bunka Kyokai [Rural Culture Association]) (in Japanese), Tavares, Carlos Eduardo Cruz (2005) Análise da competitividade da cadeia produtiva da soja em Mato Grosso. *Revista da Política Agrícola*, 14(3), pp. 75–87.

USDA (2001) *Agriculture in Brazil and Argentina: Developments and Prospects for Major Field Crops.*

USDA (2003) *Brazil: Future agricultural expansion potential underrated*, http://www.fas.usda.gov/pecad2/highlights/2003/01/Ag_Expansion/, date accessed September 22, 2010.

Valdes, Constanza (2006) 'Brazil's booming agriculture faces obstacles.' *Amber Waves*, November 1.

Warnken, Philip H. (2002) *The development and growth of the soybean industry in Brazil* (Hoboken (NJ): Wiley).

World Bank (2009) 'Awakening Africa's sleeping giant: prospects for commercial agriculture in the Guinea savannah zone and beyond,' Agriculture & Rural Development Notes.

World Bank/Brazilian Institute of Applied Economic Research (2011) *Bridging the Atlantic Brazil and Sub-Saharan Africa: South–south partnering for growth*.

World Wildlife Fund-UK (2011) *Soya and the Cerrado: Brazil's forgotten jewel*.

Yoshii, Kazuhiro, A.J. Camargo, and Álvaro Luiz Orioli (2000) *Monitoramento ambiental nos projetos agrícolas do Prodecer* (Planaltina: EMBRAPA Cerrados).

# 4

# The Impact of Cerrado Development: Stable Food Supply and Socially Inclusive Development with Value Chains

*Akio Hosono and Yutaka Hongo*

## Introduction

The extraordinary development of Cerrado agriculture led to a variety of results. Section 4.1 will discuss its contribution to the stability of the food supply, especially in terms of soybeans and corn, as well as the improvement of global food security. At the same time, Cerrado agriculture enabled the development of food value chains both inside and outside the Cerrado region and increased employment opportunities throughout the area. Inclusive and dynamic growth took place in the region, all of which will be discussed in Section 4.2. Section 4.3 will examine how this drew people to the inland of Brazil from the South and Northeast regions, reducing the excessive concentration of economic activities and population in the south of the country. This process is related to socially inclusive development in Brazil, whereby poverty has been significantly reduced and the income gap between rich and poor has considerably narrowed in the past decade, as discussed in Section 4.4. Section 4.5 will discuss how food security and lower food prices contributed to the achievement of Millennium Development Goals (MDGs) in Brazil, and Section 4.6 will examine how the expansion of value chains, critical to the creation of jobs and for socially inclusive development, was attained in the Cerrado region. Finally, Section 4.7 will present concluding remarks.

## 4.1 A stable food supply and food security: the international contribution of Cerrado agriculture

The major contributions of Cerrado agriculture have been in the areas of food supply stability and food security. These gains have been achieved despite a substantial increase in the demand for food, particularly in developing countries, due to population growth and rising income levels. In the case of soybeans, for example, China has become an importer: Its annual imports of this commodity have now reached 60 million mt, which is more than 15 times the volume imported by Japan.

Unless the global food supply improves, rapid increases in the demand for food will bring about sharp rises in food prices. This would undoubtedly have a serious effect on the many countries that rely on food imports from overseas, particularly developing countries. This is exemplified by events in 2007 and 2008. Immediately before the Lehman shock in the fall of 2008, the bubble phenomenon caused markets to speculate on food as well as oil. As a result, food prices recorded historically sharp rises and many developing countries experienced significant jumps in food prices as well as shortages. In some countries, this created circumstances that led to anti-government uprisings. Soaring prices were not only attributable to tight supply: They were partly due to speculation, and therefore temporary. Nevertheless, unless there is a sustained, long-term increase in the food supply, sharp rises in food prices will continue to occur, and this will result in a severe burden being placed on developing countries, and particularly on the poor.

The soaring prices that occurred in 2007–2008 provided a glimpse into what might happen in the near future if a stable global food supply is not secured. As a World Bank report explains, "the food price spikes in 2008 and 2011 have prevented millions of people from escaping poverty because the poor spend a large share of their income on food and because many poor farmers are net buyers of food" (World Bank, 2012, p. 5), continuing, "higher food prices have two main effects on net buyers of food: an income effect through reductions in the purchasing power of poor households, and a subnutrition effect through shifts to less nutritious food" (Ibid., p. 6).

From the 1990s onwards, Brazil, and particularly the agriculture and agri-business of the Cerrado region, have made the largest global contribution to achieving the continuous expansion of the supply of food to national and international markets, and, by extension, to ensuring a consistent worldwide food supply and price stability. For instance, in 1990, worldwide soybean production was just over 100 million mt, half of which was produced in the United States, while Brazil's production stood at just 15 million mt. By 2010, however, Brazilian production had grown to 76 million mt, accounting for almost 30 percent of the 260 million mt produced worldwide (see Chapter 3). Brazil's expanded share is even more remarkable in terms of export volume. Brazilian soybean exports as a percentage of total global exports went from 9.0 percent in 1987 to 32.2 percent in 2007. Today, Brazil's exports equal those of the United States. Back in 1987, no corn was being exported from Brazil, but 20 years later, the country accounted for 7.9 percent of global exports. Exports of soybean oil rose from 16.1 percent to 21.7 percent. Soybean meal (also called 'bean cake'), a by-product of the soybean oil extraction process, is an essential component of feed for livestock, and has come to be used in Brazil for the domestic production of meat. As a result of developments in the livestock value chains both inside and outside the Cerrado region, soybean meal has become as important as soybean oil in Brazil. For this reason, Brazilian soybean meal exports as a percentage of global exports dropped from 30.0 percent to 21.4 percent. However, Brazilian meat exports as a percentage of global exports grew remarkably: broiler chicken went from 12.9 percent to 39.6 percent and pork rose from 0.5 percent to 14.1 percent.

Agro-industry (agri-business) generated 22 percent of Brazil's GDP in 2011 (Dos Santos, 2014) in spite of the fact that agriculture alone corresponds to 5.7 percent of GDP. This industry created employment opportunities for 16 million people. The export value of agro-industry corresponds to 42.3 percent of Brazilian total exports. The net export of agri-business products has been increasing steadily since the year 2000. (Table 4.1)

*Table 4.1* Agri-business of Brazil: production and trade

| | |
|---|---|
| Increase of net export value of agri-business 2000–2013 (export value minus import value of agri-business products: 2000 = 100) | 568.1 |
| Export value of agri-business in total value of exports of Brazil 2013 (from December 2012 to November 2013) | 42.3% |
| GDP of agri-business in total GDP of Brazil 2011 | 22.15% |
| Agricultural production, processing and distribution | 15.42% |
| Input | 1.58% |
| Agriculture (production) | 3.67% |
| Industry (processing) | 5.45% |
| Distribution | 4.72% |
| Livestock production, processing, and distribution | 6.73% |
| Input | 1.03% |
| Livestock (production) | 2.71% |
| Industry (processing) | 0.87% |
| Distribution | 2.12% |

*Source:* Authors, based on CEPEA/USP, 2014, CEPEA/USP, 2012, and Ministry of Agriculture, Livestock and Supply (AgroStat Brazil).

## 4.2   Socially inclusive development of the Cerrado: expansion of the value chain and employment opportunities

As discussed above, the development of Cerrado agriculture and agribusiness has made a major contribution to global food supplies. At the same time, this process has also advanced socially inclusive development in Brazil, which makes it possible for people to escape from poverty. An expanded value chain that linked areas within the Cerrado region and also established links outside the region advanced the socially inclusive development process in a dynamic fashion. The fact that this value chain has especially gained in strength over the past ten years shows that, since the end of the 1990s, although the volume of Brazil's soybean meal exports may have plateaued, the production and export of broiler chicken and pork has continued to grow, steadily at first and then at an accelerated rate starting at the end of the 1990s. While meat exports had, prior to this, been largely ignored, they too experienced a sharp increase during this time.

Extracting oil from soybean and processing it into soybean oil and soybean meal, rather than exporting unprocessed soybeans as a primary commodity, resulted in higher added value within Brazil, and created more jobs. Furthermore, although soybean meal can be exported as is, if it is combined with corn and other ingredients to feed livestock, which is then processed into meat to be exported, its value is further enhanced and employment is again expanded. Moreover, while meat can also be exported as is, further processing will add value as well as increase employment. This applies to dairy products as well. This process of using the primary commodities from crop production, including soybean production, as a starting point to expand the value chain downstream has steadily taken place in the Cerrado since the 1990s.

In the Cerrado, it is not just the case that direct employment has expanded due to increased crop production. Soybean processing, and the production of meat, processed goods, and dairy products, have dynamically expanded the value chain, and have created employment to an extent that far surpasses direct employment from the production of agricultural commodities. There is another value chain positioned upstream, which includes inputs such as seeds, fertilizers, pesticides, and farm machinery. Furthermore, a transport and logistics network was established both for export from production sites to ports and for domestic processing, and wholesale and retail commerce mobilized an immense volume of grains and other agricultural commodities and food products throughout both domestic and international value chains or 'supply chains'. These activities generated remarkable employment opportunities throughout the Cerrado region.

As will be discussed later, the rate at which the workforce has grown in the Cerrado region (states within the Cerrado biome) is higher than in other regions of Brazil. In the Cerrado, agricultural production is based on farms of between a few and 2,000 hectares, which are not necessarily labor-intensive.[1] However, the creation of employment opportunities is significant when viewed in the context of Cerrado agri-business overall, and the region has absorbed a portion of Brazil's growing working-age population since the 1990s.

The value chain composed of soybeans, soybean products, and chicken meat in the western Bahia (BA) region provides a clear demonstration of the reality of the above process.[2] In early 1980s, at the start of PRODECER-II in the region through Japanese–Brazilian cooperation, there was no crop production, soybean processing, or livestock raising taking place whatsoever; everything had to be started from scratch. At that time, there was only a single gas station in today's Luís Eduardo Magalhães, a city situated on the western edge of the BA region. Today, Luís Eduardo Magalhães is a city with a population of 50,000 and a hub of Cerrado agriculture in the state of BA.

According to research by Professor Tetsuo Mizobe, who conducted a field survey on the upstream and downstream value chains that form the core of agri-business in this region, 3.63 million mt of soybeans were produced in the western BA region in 2010; 3.03 million mt of that was purchased by the transnational grain majors Cargill and Bunge, 2.20 million mt of which was used for soybean processing, and about 100,000 mt was used for seeds. The 1.33 million mt, comprising the remaining 730,000 mt plus an addition 600,000 mt collected by Multigrain, ADM, and Amaggi/LDC, was exported from the region overseas. From the 2.20 million mt used for processing, 1.72 million mt of soybean meal was produced, 200,000 mt of which was used for chicken feed in the region to produce 190,000 mt of broiler chicken meat. This chicken production alone, on a price-per-ton basis, raised the value of the product to approximately four times what it was originally, and the value chain created jobs for up to 74,400 workers in the western BA region, and an estimated 109,400 workers including those outside the region. These figures do not include employment through value chains other than chicken meat production (such as dairy products), nor employment related to the meat distribution and retail sectors.

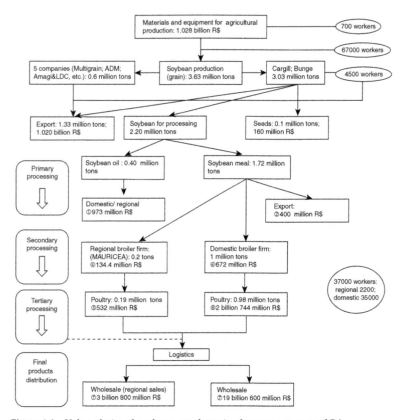

*Figure 4.1*    Value chain of soybean products in the western part of BA

*Notes:* Export value of soybeans: Export quantity 1.33 million tons * export value R$828 / ton (FOB at Port Paranagua, 2010, http:www.abi ove.com.br). (* indicates hereinafter 'multiplied by')

Seed production value: Seed production quantity * seed price R$1.6 / kg (Agrianual, 2011). Soybean oil production quantity was calculated at 18% of total processing of soybeans, and soybean meal quantity was calculated at 78% yield rate.

Soybean meal production quantity * export value R$744 / ton (FOB at Port Paranagua, 2010, http:www.abiove.com.br).

Poultry production value: Shipping amount * value at plant R$2.8 / kg (value based on an interview with MAURICEA in Bahia).

Wholesale: Sales * retail price R$20 / kg (price based on interviews with retail stores in São Paulo).

Logistics and labor absorption are excluded from the total value-added calculation due to difficulty of calculation.

Labor absorption: (1) The fertilizer company, Galvani, 500, fertilizer and agricultural machinery agents, 200 (50 agents * 4 persons each); (2) Number of workers in agricultural fields was calculated by total payroll/minimum wage; (3) Local grain companies were based on interviews; (4) Broiler companies were based on interviews.

*Source:* Authors, based on Mizobe (2014).

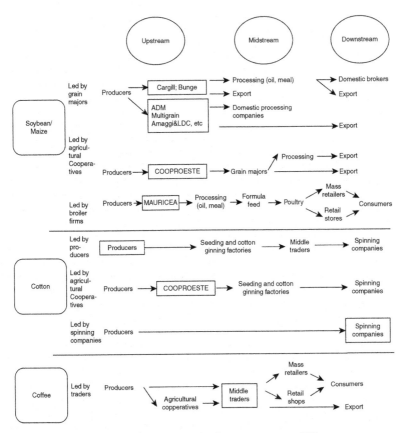

*Figure 4.2*   Value chains of main crops in the western part of BA
*Source:* Authors, based on Mizobe (2014).

Moreover, as shown above, in addition to the value chains based on soybean and chicken meat production, value chains and other mechanisms that center on cotton and coffee are also undergoing development in the western BA region.

Similarly, across the entire Cerrado region, agriculture, livestock, and processing chains, including those for beef, pork, dairy, vegetables, and fruit, continue to expand and deepen their roots. It goes without saying that the increase in added value and expanded employment these advancements have made possible far exceeds those from crop production itself.

In addition to value chains around grains, the development of labor-intensive agriculture, such as vegetables and fruit, has contributed to increasing employment in the Cerrado. The growing of vegetables by small farmers deserves special mention, because the initiative of vegetable farming in the Cerrado around Brasilia started around the same time as the construction of the new capital city, in the late 1950s.

As Brasilia was being built in the late 1950s, then President Juscelino Kubitschek invited farmers of Japanese descent near the city of São Paulo to develop the area around the new capital. The city needed agricultural outputs to cope with the influx of people the construction had brought. In 1959, the settlement began in the Vargem Bonita neighborhood. Each plot was small: 4 ha. Half a century after a difficult process of reclamation, crops are now grown in greenhouses.

Many farmers of Japanese descent continue to farm in the area, and many small farmers have succeeded like Japanese–Brazilian farmers in the Cerrado. For example, the Brazilian Institute of Agrarian Reform (INCRA) settlement was established in the western part of Brasilia, in two neighborhoods named Alexandre Gusmão and Brazlândia, both of which are dominated by small farmers dedicated to vegetable growing. Over 3,000 households farm the land. Most of the fields are less than 5 hectares and the work is performed by families. In the INCRA settlement, orchards are also common, and fruit processing plants were constructed to strengthen fruit value chains. Once it had become possible to process and store crops, farmers could offer a stable production volume and achieve continuous sales; this led to income stabilization.

These small farmers receive support from extensionists in the Federal District Technical Assistance and Rural Extension Corporation (EMATER-DF), a state agricultural extension entity. EMATER provides economic assistance and practical support to farmers in choosing crop varieties and using chemicals and fertilizers. In order to support vegetable production by small farmers, EMBRAPA's research center specializing in vegetables (EMBRAPA Vegetables) has been making important technological developments for more than 20 years. JICA has cooperated with the center throughout this process, sending experts to Brazil and inviting Brazilian professionals to Japan.As discussed above, Cerrado agriculture is highly diversified today, and farms of different sizes coexist. Their size and employment effects generally depend on the types of products they produce.

## 4.3   Population migration toward inland areas along with the development of Cerrado agriculture

The process outlined above is what has brought about the increases in employment and population in the Cerrado region over the past

20–30 years. As a result of Brazil's long history of development as a part of the New World, the country's population has always been concentrated along the coast, especailly in large cities, leaving inland areas sparsely populated. In fact, the word for 'inland' or 'the interior' in Brazil implies an outlying and remote region. Cerrado agricultural development stimulated migration, attracting people from the coast to inland cities and villages.

Table 4.2 shows the change in the population and distribution ratios for each region of Brazil, every ten years beginning in 1970, which was prior to the start of Cerrado agriculture.[3] According to this table, the population ratio of the Central-West region where a large part of the Cerrado is located, increased from 4.9 percent to 7.4 percent from 1970 to 2010.

While Brazil's population doubled in the 40 years from 1970 to 2010, the ratio gradually declined in the heavily populated northeastern, southeastern, and southern regions, while the ratio in the so-called inland regions with low population density, the vast Central-West and North, gradually increased. The population increased five-fold during this period in MT (Central-West), which produces the largest volume of soybeans in the Cerrado region.

Brazil began to adopt measures to promote inland areas starting in the 1960s, as illustrated by the transfer of the capital from Rio de Janeiro to Brasilia in 1960. For the Cerrado's development, the federal government exercised leadership in promoting investment in research, logistics, infrastructure, and agricultural development starting in the mid-1970s, as discussed in Chapters 1 and 2. Private sector entities utilized the

*Table 4.2* Change in total population and population ratio (%), by region

| Region | 1970 | 1980 | 1991(*) | 2000 | 2010 |
|---|---|---|---|---|---|
| Total population (unit: 1,000) | 94,508 | 121,150 | 146,917 | 169,590 | 190,755 |
| North (45% of the overall land area of the country) | 4.4 | 5.6 | 7.0 | 7.6 | 8.3 |
| Mid-west (19%) | 4.9 | 5.8 | 6.4 | 6.8 | 7.4 |
| Northeast (18%) | 30.3 | 29.2 | 28.9 | 28.1 | 27.8 |
| Southeast (11%) | 42.7 | 43.4 | 42.7 | 42.6 | 42.1 |
| South (7%) | 17.7 | 16.0 | 15.1 | 14.8 | 14.4 |

*Note:* * Survey for 1990 was conducted in 1991.

*Source:* Authors, based on IBGE statistics (2011).

outcomes of these projects and made inroads into the area, which promoted further economic activity. These factors led to migration from other areas of the country into the Cerrado region.

Research carried out using the National Household Sample Survey (PNAD) conducted by the Brazilian Institute of Geography and Statistics (IBGE) (Nishijima and Hamaguchi, 2011) pointed out that reductions in regional income disparities and population shifts were becoming increasingly evident in the past decades, mentioning a declining trend in the supremacy enjoyed by the country's southeast. The researchers concluded: "The results of the 2008 PNAD survey suggest that an 'economy of accumulation', in which highly educated, skilled labor was being drawn from the southeast, was taking place in inland regions that were once considered backward, and indicate the possibility that in the long-term, the overconcentration in the southeast that formerly characterized Brazil will shift, transforming the country into a more decentralized one. This could signal a critical turning point for the Brazilian economy" (Ibid., p. 111; translation by the authors.).

## 4.4   The increase in employment in the Cerrado that contributed to socially inclusive development in Brazil

The influx of workers attracted by greater employment opportunities in the Cerrado is the crucial factor behind the phenomenon that could be a turning point in Brazil's economic development. The population growth rate in the Cerrado region (in states that include part of the Cerrado biome) has consistently outpaced that of non-Cerrado regions. Analysis that separates Brazil's workers and demographics into categories for Cerrado and non-Cerrado regions makes this clear. According to the 2010 PNAD survey, the number of workers nationwide increased from 76.16 million in 2001 to 92.69 million in 2010. A breakdown of these figures shows that while the number of workers in the Cerrado region rose by 27 percent, from 29.7 million to 37.7 million, the number of workers in non-Cerrado regions rose by 18 percent, from 46.5 million to 55.0 million.

According to the population census, the population in the Cerrado region more than doubled (2.12 times) over the 40-year period from 1970 to 2010, going from 35.8 million to 76.0 million. On the other hand, the population in non-Cerrado regions increased 1.96 times, going from 58.6 million to 115 million over the same period. It is important to note that the absorption of population by many of the huge cities in the southeast, such as Rio de Janeiro, São Paulo, Curitiba, and Porto Alegre, remained strong throughout this period; and yet, despite the fact that these huge cities in non-Cerrado regions swelled and their populations became increasingly concentrated, the inner Cerrado region achieved a population growth rate which surpassed that of non-Cerrado regions. Another point to note is the increase in the number of farms in the Cerrado region. According to the agricultural census, the number of farms in the region continued to grow at a rate of 5.1 percent per year from 1970 to 2006. At the same time, in non-Cerrado regions, the number of farms declined at a rate of 8.6 percent per year. In the Cerrado region, the proportion of employment in rural areas was 36.3 percent in 2009, but was just 21.2 percent for non-Cerrado regions.

From the 1990s into the present century, Brazil has been going through a period on the verge of the demographic bonus effect. For the 30-year period from 2010 to 2040, Brazil will experience the demographic bonus effect. While the Brazilian population grew at a rate of 3.3 percent per year in the 1960s, followed by 2.0 percent in the 1970s, 2.8 percent in the 1980s, and 1.7 percent in the 1990s, it can be said that expanded employment opportunities in the Cerrado continues at least partially absorbing this increasing population, and particularly the productive-age population.

According to studies by Arnold and Jalles (2014) and the OECD (2011), Brazil is regarded as the only major present-day emerging nation that is improving its distribution of income. Brazil has traditionally had a high Gini coefficient and has been known as a country with high income disparities. While a high Gini coefficient means that there is a lot of room for inequality to decrease, at the same time, like other developing nations, Brazil ran the risk of growth going hand in hand with increasing disparity. However, Brazil managed to improve its distribution of income while steadily achieving growth. Arnold and Jalles (2014, p.5) showed that the Gini coefficient decreased from around 0.60 in late 1990s to less than 0.53 in 2012 in Brazil.

Many countries have failed to achieve socially inclusive development, in which all people participate in growth and at the same time all people benefit from growth.[4] However, two thirds of the countries in the world experienced an increase in income inequality despite solid growth between 1990 and 2005 (United Nations 2012, p.5)

As is well known, significant regional disparities have existed in Brazil since colonial times. There is a disparity between the northeast, which mobilized large numbers of slaves as its manpower, and the São Paulo-based southeast, which flourished due to its coffee industry and subsequent industrialization, as well as a disparity between coastal regions and inland regions. In Brazil, the term *nordeste* (northeast in Protuguese) is used as a synonym for poverty, an association that continues to this day.

By narrowing regional disparities between coastal and inland regions and expanding employment opportunities in the Cerrado region, Cerrado development is thought to have made considerable contributions towards conquering poverty and regional disparities. Steady progress has been made over the past 40 years, starting in the mid-1970s, and has been particularly notable over the last 20 years, starting from the 1990s, during which time the agricultural and livestock processing value chain has expanded. However, six of the nine northeastern states of Brazil, the country's poorest area, fall outside the Cerrado region: in northeastern Brazil, only BA, MA, and PI include Cerrado biome zones. Nevertheless, even the remaining six northeastern states have benefited from Cerrado agriculture, because many people migrated from these six states into the Cerrado region (Figure 4.3).

The consolidation of the Real Economic Plan (Plano Real) under President Fernando Henrique Cardoso started a new era in Brazilian social and economic development. Most of the results obtained in the following administration would not have been possible without the gains obtained from the Plano Real.

President Luiz Inacio 'Lula' da Silva, who succeeded Cardoso, was born into a poor farming family in the northeast. He was only able to attend compulsory education, and from childhood he worked as a street peddler and held other jobs before committing himself to the labor movement. President Lula went to extraordinary lengths in his efforts to rid Brazil of famine and poverty, and his administration actively tackled social policies and measures to fight poverty in particular.

Under the Lula administration, efforts to reduce poverty and narrow disparities yielded significant results and attracted worldwide attention. The greatest of these efforts was a direct financial assistance program for poor families dubbed the *Bolsa Família* (Family Allowance), which is still paid to 13.4 million households. Payment of the allowance is conditional on the family's children attending school and being immunized, and it aims to alleviate poverty, improve school attendance, and reduce infant mortality rates. This kind of approach to reducing poverty has recently come to be referred to as conditional cash transfer, or CCT, and Brazil, along with Mexico, is a global pioneer in this measure against poverty.

Brazil's *Bolsa Família* was originally a policy implemented as part of the Lula administration's *Fame Zero* ('Zero Hunger') program. As a

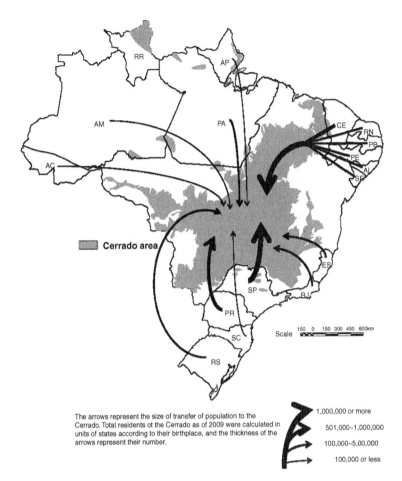

*Figure 4.3*  Transfer of population to the Cerrado
*Source:* Authors, based on IBGE statistics.

full-scale initiative aiming to assist the poor, whom successive Brazilian administrations had failed to adequately address, *Fame Zero* was a policy that best reflected the characteristics of the Lula administration. Various other programs were implemented, including the *ProUni* program, which aimed to improve literacy rates and enable those in lower income groups to go to university. These efforts yielded marked results that continue to this day.

Considered from a longer-term perspective, the control of inflation, which began before the Lula administration (during that of Cardoso), was hugely beneficial for the poor. Combined with the later rise in the minimum wage and higher rates of growth, measures to fight poverty such as the *Bolsa Família* proved successful. The population with household income per capita below the extreme poverty line (US$ 2 per day) decreased from around 30 percent in mid 1990s to near 5 percent in 2012. (Arnold and Jalles, 2014, p. 5)

In this way, Brazil which had been known to have one of the world's highest levels of income disparity, managed to gradually narrow such gaps. The Lula administration can be regarded as having produced remarkable achievements both in its social policies and in its measures to fight poverty and narrow income disparity.

Thus it can be seen that several factors came together to achieve inclusive development in Brazil from the 1990s onward. It is beyond doubt that social policies focused on direct measures to fight poverty have been successful; in addition, the macro-economic policies that curbed inflation, and the sustained growth that took place under these policies were also important.

Furthermore, the growth of Cerrado agriculture could have contributed to the above-mentioned process of inclusive development in two principal ways. First, poor people's access to affordable food with improved nutrition could have been improved substantially as a result of increased food security and decreases in the real prices of food, due both to the increases in food production and the conversion of Brazil from a net importer of food into a large net exporter. It could have produced a positive impact on the reduction of poverty, the nutrition of children, and the reduction in child mortality rates, which are crucial indicators of Millenium Development Goals (MDGs).

Second, in the inland Cerrado region, which had traditionally lagged far behind the coastal regions, a value chain based on groundbreaking agricultural and livestock processing was developed, and an employment growth rate that outpaced non-Cerrado regions was achieved. These accomplishments are thought to have had a significant effect on narrowing the disparities between coastal and inland regions.

Given this combination of factors, the following sections will discuss how food security and a lower real price of food contributed to the achievement of MDGs in Brazil, and how the expansion of value chains, critical for employment generation and inclusive development, was attained in the Cerrado region.

## 4.5    Food, nutrition, and achievement of MDGs in Brazil

GDP growth does not necessarily lead to poverty reduction. Furthermore, even if the percentage of the population living on less than US$1.25 per day (purchasing parity base) decreased, their nutrition would not improve if food price hikes occurred – as has happened, for example, in Sub-Saharan Africa, where, in spite of GDP growth, the share of the population living in extreme poverty increased six percentage points to 58 percent from 1981 to 1999 (UNDP, 2012, p. 19). Moreover, chronic malnutrition, measured by the share of preschool children who are stunted, is estimated to have fallen only two percentage points (from 43 percent to 41 percent) between 1990 and 2010. In Asia (excluding Japan) the percentage of such children declined by more than 25 percent from 2000 to 2010 (Ibid., pp. 19–20).

In Brazil, economic growth was accompanied not only by a substantial reduction in poverty, but also by improvements in nutrition. The percentage of the population surviving on less than US$1.25 per day had fallen from 25.6 percent in 1990 to 6.1 percent in 2007, when the MDG goal of poverty reduction was achieved in Brazil, after the rate fell to less than one-fourth of what it was in 1990. The rate in 2011 was as low as 2.3 percent (Ministry of Social Development and Hunger Alleviation, 2010). During this time, malnutrition was also reduced: in 1996, 4.2 percent of Brazilian children under four years of age were below the expected weight for their age. In 2006, that figure had dropped to 1.8 percent. (Ministry of Social Development and Hunger Alleviation, 2010)

It is well known that some of the MDGs are closely linked to food and nutrition, particularly child mortality (MDG4) and maternal mortality (MDG5) rates. Globally, progress regarding these goals is lagging among MDGs. Out of the 144 countries being monitored, 105 are not expected to reach MDG4, and 94 are off-track on MDG5. In Brazil, child mortality per thousand live births had been reduced from 53.7 in 1990 to 22.8 in 2008, and the country is on track to achieve the target of 17.9 before 2015. Regarding MDG5, from 140 deaths per 100,000 live births in 1990, the ratio had sharply declined to 75 deaths in 2007 (Ministry of Social Development and Hunger Alleviation, 2010).

In the food–nutrition–poverty nexus, the key factor appears to be food security and food prices. "Poverty typically rises initially with higher food prices, because the supply response to higher prices takes time to materialize and many poor (farm) households are net food buyers, so higher food prices lower their real income" (World Bank, 2012, p. 3). As cited above, higher food prices have two main effects on the deterioration of poverty: "an income effect through reductions in purchasing power of poor households, and a substitution effect through shifts to less nutritious food" (Ibid., p. 6).

As has been mentioned, the level of malnutrition is closely related to some critical MDGs:

> Child malnutrition accounts for more than a third of under-five mortality (MDG4) and malnutrition during pregnancy, for more than a fifth of maternal mortality (MDG5). Furthermore, conditions during early life (from conception to two years of age) provide the foundations for adult human capital and vicious circles of malnutrition, poor health, and impaired cognitive development set children on lower, often irreversible, development paths. (Ibid., p. 6)

From this perspective, the continuous trend of lower food prices in Brazil is highly significant; the real price of food in Brazil decreased substantially during the 25 years between 1975 and 2000 (De Barros et al., 2001). The beneficiaries of this decline in the price of basic foods are low-income consumers. Brazil's poverty reduction, which came with the income increases among the low-income population, took place together with a substantial decline in the real price of food, which therefore improved nutrition and achieved nutrition-related MDGs, especially the goal concerned with the child mortality rate (MDG4).

## 4.6 Factors enabling the development of dynamic food value chains in the Cerrado

A number of requirements can be mentioned for the formation and expansion of a value chain. Related to supply conditions, these include having a competitive raw material, the technology to process that material, machinery that enables processing, and investment assets, and a capable labor force. Among these factors, competitive raw material is considered crucial. A recent study on agri-business discusses how inputs are a binding constraint for competitive value chain formation in Sub-Saharan Africa (Dinh et al., 2012, p. 57).[5] There are also demand conditions comprising internal and external markets through which processed goods can be sold. In the case of the Cerrado, the most notable of the above-mentioned demand and supply factors were the high competitiveness of the raw materials, the abundant labor force, and the existence of processing technologies.

The highly competitive raw materials produced in the Cerrado, particularly in the agricultural and livestock processing business, promised high profit margins. Items produced in the Cerrado region, namely crops such as soybeans, corn, cotton, and coffee, are typical international commodities, and their prices are determined by the international market. The prices determined at the Chicago Board of Trade (CBOT) usually serve as the standard for soybeans and corn, and the CBOT has the greatest influence over worldwide crop prices. Suppose, for example, that exports from Brazil are bound for Amsterdam. If delivery took place at a silo at the place of production in the Cerrado, the price paid to the producer would be the international price minus the transportation costs from the silo to the Brazilian export port and then to Amsterdam. As a result, if the Amsterdam import price was US$436 per mt, the price of silo delivery in the western BA region of the Brazilian Cerrado would be the remaining US$302 after transportation costs to Amsterdam were deducted.

Although this is a factor that introduces a disadvantage in the export of soybeans from the Cerrado region as an international commodity (that is, as a primary commodity), the Cerrado region finds itself under such adverse conditions due to factors that cannot be resolved in a short time. Specifically, these factors are the distance to the coast, which is a preexisting factor, and delays in the upgrading of infrastructure. It should be noted that the Cerrado value chain turns these adverse factors into an asset. That is to say, the trading prices of raw materials (soybeans, corn, and so on) inside the Cerrado region are far lower than their corresponding international prices, and they have a significant advantage compared to importing the same goods to produce agricultural, livestock, or processed goods. These conditions became possible due to the expansion of crop production in the Cerrado.

Agricultural and livestock processing technologies has been developed over a long period in the southeastern agricultural zone, and there has been also access to an extensive labor force all over Brazil. By having this labor force handle processing and production, the Cerrado region managed to build a competitive agricultural, livestock, and processing value chain. Put another way, the high costs of transportation from inland areas to ports as well as high port fees (these are part of what is known as the 'Brazil cost') remain the same, and various initiatives such as improvements to infrastructure are required to reduce these costs. However, it will take time to improve these circumstances in the future.

Based on this situation, the business of agricultural and livestock processing and the people and companies associated with it developed a robust strategy in the Cerrado, which has enabled high-value-added processed goods to be produced and employment to be expanded inside Brazil.

There are many organizations participating in the value chain. In the western BA region, the Western Bahian Technology and Research Center (CPTO) coordinated with a branch office of EMBRAPA on testing, research, and technological development. The resulting technologies were provided to producers through regional producer support organizations such as the Agricultural Cooperative of Western Bahia (COOPROESTE). In addition to soybean research, the CPTO developed pest-resistant varieties of key crops such as cotton, promoted the introduction of irrigation systems and new crops such as sunflowers, and engaged in technological development aimed at improving productivity (Mizobe, 2014).

At the same time, GTCs and lower-ranked crop trading companies (eight companies in all) provided farming loans and purchased agricultural production; agri-business companies sold agricultural inputs such as machinery, fertilizers, and pesticides and provided related technical support, and agricultural consultants also became involved. In this way, an articulated structure of organizations (often called a 'cluster'), with the CPTO at the center, was created to engage in innovation and its dissemination. These organizations became hubs in the value chain, and more clusters of such hubs engaging in innovation as part of a national or sectorial system of innovation are being formed (Figure 4.4).

The western BA region mentioned in this chapter produces 98 percent of all soybeans, 70 percent of all corn, and 100 percent of all cotton in the state of BA. The development of Cerrado agriculture and agribusiness in this area started with the pilot project of the second phase of the PRODECER program, conducted here through Japanese–Brazilian cooperation from 1985 to 1990. In 1980, the soybean crop area in the region was just 1,000 ha, which grew to more than a million hectares by 2010. The total cultivated area in the region increased from 470,000 ha in 1992 to 1,840,000 ha in 2010. The growth rate of total agricultural output during this period reached 14.6 percent per year.

As a result of this process, major changes were seen among the domestic and foreign companies operating in the Cerrado. First of all, related to the expansion of the value chain, the purchase, transportation, and export of crops, which had been conducted by GTCs in what was largely an oligopoly, gradually gave way. For instance, in western BA, the two grain majors, Cargill and Bunge, held a virtual monopoly over

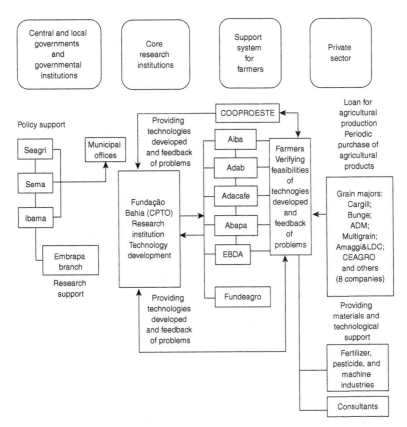

*Figure 4.4* Connections between government, agri-business, and farmers in the western part of Bahia

*Source:* Authors, based on Mizobe (2014).

the handling of soybeans until the mid-2000s, but in recent years, ADM, Multigrain, the Amaggi Group, and other crop trading companies have entered the market. In addition, companies involved in feed and livestock production and meat processing moved into the Cerrado. Many of these were local companies. With the diversification of agricultural production in the Cerrado, the participation of companies that deal with new agricultural products of expanding value chain is also significant. In western BA, this is noticeable with regard to cotton. Furthermore, there also striking efforts being made by companies dealing with the upstream components of the value chain, namely fertilizer, seeds, pesticides, and farming machinery.

In the Cerrado, the western BA region is just one example of an area where PRODECER programs have sparked development. It is safe to say that development similar to that of the western BA region has been achieved in many areas in which a PRODECER program has been pursued. Based on municipal-level human development indicators, it can be confirmed that many of the sites of PRODECER programs have developed as major urban areas, produced the high-levels of life satisfaction in each of the states where these sites are located, and above all increased the quality of life of the local people. More than anything, this speaks volumes about the impact of PRODECER and other programs implemented in the Cerrado. (See Figure 4.5)

In his research into the development of the Cerrado production structure from 1975 onward, Dr. Bernardo Palhares Campolina Diniz of the Federal University of Tocantins emphasizes that these high-levels of satisfaction are due to the programs that were put into action, namely production-oriented initiatives such as infrastructure development, public and private settlement, POLOCENTRO and PRODECER, as well as programs that aimed to develop the Cerrado on a systematic basis. His research also focuses on the livestock, soybean processing, sugar, and alcohol industries, and analyzes population growth and the makeup of urban networks explaining the development of value chains and their impact in the state of Tocantins. (Campolina, 2006).

## 4.7   Concluding remarks

We can conclude that Cerrado development through Japanese–Brazilian cooperation was one of the main factors that triggered or catalyzed the expansion of the value chains based around Cerrado agricultural and livestock processing businesses. During this period PRODECER and other initiatives, both public and private, led to the formation of clusters that fulfill core roles in the formation of food value chains. The value chains, together with agriculture, have created many employment opportunities surpassing those in non-Cerrado regions, produced migration to inland states, and contributed to socially inclusive development with expanded employment, reduced poverty, malnutrition, child and maternal mortality, improving the quality of life of the local people. The discussion on long-term social impact of Cerrado development deserves a further in-depth research.

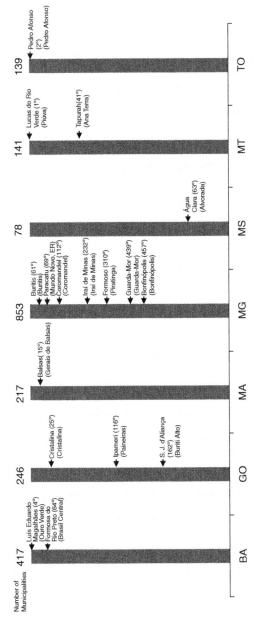

*Figure 4.5* Ranking of municipalities that participated in PRODECER in terms of employment and income (2009)

*Source:* Authors, based on Firjan (Federation of Industries of the State of Rio de Janeiro) data.

## Notes

1. Farm size depends on types of agricultural products. Today, farm size is small for horticulture (a few hectares) and large for grain production (500 to 2,000 ha).
2. Field surveys on the agriculture and agro-industrial value chain in western Bahia were made by the authors and Prof. Mizobe. For further details see Mizobe (2014).
3. Brazil is a federal nation, consisting of 26 states and the Federal District of Brasilia. Each state is divided into counties (*municipios*), and there are a total of 5,565 counties in the country. At the same time, the government and IBGE divided the country into five regions (*regiões*) based on their regional characteristics, and the public has widely used these categories. These are: (1) the Northern region (where the major vegetation is the Amazon tropical rainforest); (2) the Mid-west (the Cerrado); (3) the Northeast (semi-arid); (4) the Southeast (the four advanced states); and (5) the South (agricultural). IBGE has compiled demographic survey results at three levels (county, state, and region). (See the frontispiece for a map showing the regions.)
4. For discussion on inclusive development, see Kozuka (2013), Hosono (2015a) and Hosono (2015b).
5. This study discusses the potential of light manufacturing, including agri-business, and emphasizes that without the ability to acquire a large volume of diverse inputs at competitive prices, with consistently high quality, and at short notice, Africa-based firms cannot hope to achieve competitiveness in the international market. In the case of the Cerrado agri-business value chain, the inputs met these conditions and strong competitiveness was achieved.

## Bibliography

Agra FNP Research (2011) *Agriculture in Brazil Yearbook 2011: Brazil Agrianual.*(Sao Paulo: FNP)

Arnold, Jens and Joao Jalles (2014) 'Dividing the Pie in Brazil: Income Distribution, Social Policies and the New Middle Class,' OECD Economics Department Working Papers, (Paris: OECD).

Campolina, Bernardo Palhares (2006) 'O grande cerrado do brasil central: geopolítica e economía,' Thesis (São Paulo: University of São Paulo).

Centro de Estudios Avanzados em Economia Aplicada (ESARQ/USP) (2012) *PIB do Agronegocio* Centro de Estudios Avanzados em Economia Aplicada (CEPEA/ESALQ/USP) 'Faturamento e volume exportado do agronegocio brasileiro sao recordes en 2013.' (Sao Paulo: CEPEA/ESALQ/USP).

De Barros, Mendonca, Jose Roberto, Juarez Alexandre Baldini Rizzieri, and Paulo Piccheti (2001) *Os efeitos da pesquisa agricola para o consumidor: Relatorio final* (São Paulo: Fundacao Instituto de Pesquisas Economicas (FIPE)).

Dinh, Hinh T., Vincent Palmade, Vadana Chandra, and Frances Cossar (2012) *Light manufacturing in Africa: Targeted policies to enhance private investment and create jobs* (Washington, D.C.: World Bank).

Dos Santos, Gesmar Rosa (2014) 'Agroindustria no Brasil: Um olhar sobre indicadores de porte e expancao regional,' IPEA, *Radar No.31* Brasilia: IPEA IBGE, *Indicadores IBGE Contas Nacionais Trimestrais* (Rio de Janeiro: IBGE).

Hosono, Akio (2015a) 'Industrial transformation and quality of growth', Haddad, Lawrence, Hiroshi Kato and Nicolas Meisel (eds.) (2015). *Growth is Dead, Long Live Growth: The Quality of Economic Growth and Why it Matters*. (Tokyo: JICA Research Institute)

Hosono, Akio (2015b) 'Transforming for jobs and inclusive growth: Strategies for Sub-Saharan countries' Chandy, Laurence, Hiroshi Kato and Homi Kharas (eds.) (2015) *The Last Mile in Ending Extreme Poverty* (Washington, D.C.: Brookings Institution)

Kozuka, Eiji. 2014. 'Inclusive development: Definition and principles for the post-2015 development agenda.' Hiroshi Kato, ed., 2014. *Perspectives on the Post-2015 Development Agenda*. (Tokyo: JICA Research Institute).

Ministry of Agriculture, Livestock and Supply (2014) *AgroStat Brazil*, http://sistemasweb.agricultura.gov.br/pages/AGROSTAT.html, date accessed May 23, 2014.

Ministry of Social Development and Hunger Alleviation (2010) *New extreme poverty estimates* (Brasilia: Ministry of Social Development and Hunger Alleviation).

Mizobe, Tetsuo (2014) 'Soybean products value chain in the Brazilian Cerrados: value expansion process from soybean products to chicken.' *Journal of Agricultural Development Studies*, 25(1)47–53.

Nishijima, Shoji and Nobuaki Hamaguchi (2011) '*Burajiruni okeru Keizaijiyuka no Jisshoukenkyu*' [An empirical study on economic liberalization in Brazil] (Kobe: Kobe University Research Institute for Economics and Business Administration).

OECD (2011) *Growth, employment and inequality in Brazil, China, India and South Africa: An overview* (Paris: OECD)

United Nations (2012) *Addressing Inequalities: The Heart of The Post-2015 Agenda and the Future We Want All*, (New York: UN System Task Team on the Post-2015 UN Development Agenda, United Nations)

UNDP (2012) *Africa human development report 2012: Towards a food secure future* (New York: UNDP).

World Bank (2012) *Global monitoring report 2012: Food prices, nutrition, and the Millennium Development Goals* (Washington, D.C.: World Bank).

# 5
# Cerrado Agriculture and the Environment

*Akio Hosono and Yutaka Hongo*

## Introduction

This chapter will discuss Cerrado agriculture from an ecological and environmental perspective and analyze the sustainable development of agriculture and livestock in the tropical savannah biome. The first section presents an overview of biodiversity and environmental conservation efforts based in the Cerrado, especially in the 20 years following the Rio Summit in 1992 ('Rio 92,' United Nations Conference on Environment and Development). Environmental conservation initiatives in the Japanese–Brazilian Cooperation Program for Cerrados Development (PRODECER) are then discussed in Section 5.2. It is worth noting that although this bilateral cooperation was conceived nearly 20 years prior to Rio 92, Brazil and Japan addressed environmental conservation as early as the first phase of PRODECER. In the following Section 5.3, the pattern of land use in the Cerrado is reviewed, drawing on Landsat satellite image information. From this analysis, it is estimated that 61 percent of the Cerrado remained as natural vegetation after more than two decades of intensive agricultural development.

It was in 2003, when over 20 years of PRODECER was coming to an end, that a transnational grain trade company (GTC), Cargill, constructed a soybean export terminal port and 60,000-ton storage silo in the city of Santarém, located midstream along the Amazon River. Section 5.4 of this chapter analyzes how the 'Soy Moratorium' – together with concerted efforts by the major stakeholders – limited the destruction of tropical rainforest in the Amazon. Section 5.5 discusses new Japanese–Brazilian cooperation efforts to conserve the environment, in both the Cerrado and the Amazon, including the Advanced Land Observing Satellite (ALOS) Image Utilization Project, which has been hugely effective in combatting illegal logging. Section 5.6 concludes the chapter, focusing on possible future challenges related to sustainable agricultural development in the Cerrado.

## 5.1 Biodiversity and environmental conservation efforts in the Cerrado

According to the Brazilian Ministry of the Environment (Ministerio do Meio Ambiente, MMA), the Cerrado biome is regarded as the second most biologically diverse region of Brazil, after the Amazon tropical rainforest biome. In terms of flora, the region is home to some 12,356 native plant species (roughly 44 percent of which are endemic to the Cerrado region), and its fauna includes 320,000 species (90,000 of which are insects). In particular, the transitional zones between the Cerrado biome and the adjacent Amazon tropical rainforest biome, and between the Cerrado biome and the Caatinga biome (a semi-arid zone in the northeast), have a wide variety of valuable endemic species. The Cerrado has become known as the savanna zone with the world's richest biodiversity (Ministerio do Meio Ambiente, 2010).

Plants in the Cerrado cope with the unique stresses of extreme nutrient shortages, high soil acidity, and high aluminum saturation, and are believed to have evolved to protect themselves against damage from ants and wildfires, making them a valuable genetic resource. In 1998, the NPO Conservation International, based Arlington, Virginia, designated the Cerrado region as a biodiversity hotspot.[1] The organization pointed out that original vegetation in the Cerrado region had decreased and become segmented due to agricultural development, and that the biodiversity of the region may be under threat.

In addition, the Cerrado region extends over Brazil's central highland plateau (*Planalto central*). This is an important location crossed by Brazil's three major rivers: the Amazon, Paraná (La Plata tributary), and the São Francisco (upper catchment area), which traverses the semi-arid region longitudinally from north to south and is the longest river in northeastern Brazil. Brazil's Forest Law (enacted in 1965) designated the portion of owned land for environment conservation, but that portion differs depending on the vegetation (biome). The percentage of the Cerrado region to be kept as 'legal reserve' is 20 percent.[2]

In the 1980s, worldwide concern over global warming and environmental conservation gradually mounted, almost in tandem with the development of Cerrado agriculture. In 1981, Brazil enacted its own basic environment law in the form of the National Environmental Policy Act; the spirit of this law was also incorporated into Brazil's new constitution during the shift to civilian rule in 1988. In 1989, the Brazilian Institute of Environment and Renewable Natural Resources (IBAMA) was founded to enforce Brazil's environmental policies.

Then, in 1992, Brazil hosted the world's first United Nations Conference on Environment and Development (UNCED), commonly known as Rio 92, since it was held in Rio de Janeiro, and was an active participant in the proceedings. In June 2012, the conference was again held in Rio de Janeiro (called 'Rio+20') to mark the 20$^{th}$ anniversary of the UNCED. A significant focus of the conference was the results of initiatives regarding environmental conservation and development implemented around the world over the past 20 years.

During this 20-year period, Brazil conducted a series of broad and varied initiatives aimed at environmental conservation. In the Cerrado region, the government especially pursued a balance between Cerrado agricultural development policies and environmental conservation policies.

In 1998, Brazil enacted the Environmental Crimes Law; and in 2000 the Forest Code was amended to enforce more strictly legal reserve percentages on landowners, and to enable the trading of reserve land with the land of other forested land owners. Also in 2000, the National System of Nature Conservation Units (SNUC) was established. This system was designed to organize categories for native reserves and to protect and restore the biodiversity found in their ecosystems.

From 2002 onwards, the Brazilian government further strengthened its efforts to protect the environment and Brazil's ecosystems. The establishment of preservation districts (described below), the Environmental Conservation Expansion Program, and the Action Plan for Prevention and Control of Deforestation were particularly important aspects of these efforts.

At the beginning of this period, in March 2002, the National Environmental Council (Conselho Nacional do Meio Ambiente, CONAMA)-303 environmental resolution designated water resource recharging areas, sloping areas susceptible to soil erosion, valley and floodplain forest areas, and the like as Permanent Protection Areas (APPs), even when they were within privately owned land. In addition, human disturbance of original vegetation was banned.

In 2009, the National Policy on Climate Change was established (Law No. 12187). This required the formulation of Action Plans for the Prevention and Control of Deforestation on a biome basis. In September of the following year, the Action Plan for Prevention and Control of Deforestation and Wildfires in the Cerrado (hereafter, PP Cerrado) was announced.

Table 5.1 Protected areas in the Cerrado (by unit and regulatory body)

| Protection Unit | Federal protection area | | | State protection area | | | Federal and State protection area | | |
|---|---|---|---|---|---|---|---|---|---|
| | Number | Area (km²) | % | Number | Area (km²) | % | Number | Area (km²) | % |
| **Strictly protected areas** | 22 | 41,166 | 2.02 | 86 | 16,943 | 0.84 | 108 | 58,11 | 2.85 |
| Ecological station | 5 | 10,927 | 0.54 | 23 | 528 | 0.03 | 28 | 11,455 | 0.56 |
| National Monument | 0 | 0 | 0.00 | 4 | 296 | 0.01 | 4 | 296 | 0.01 |
| National park | 15 | 28,925 | 1.42 | 50 | 14,820 | 0.73 | 65 | 43,745 | 2.15 |
| Forest ecosystem reserve | 1 | 1,280 | 0.06 | 3 | 1,188 | 0.06 | 4 | 2,469 | 0.12 |
| Ecosystem reserve | 1 | 34 | 0.00 | 6 | 111 | 0.01 | 7 | 146 | 0.01 |
| **Sustainable use areas** | 145 | 18,731 | 0.91 | 143 | 90,935 | 4.47 | 288 | 109,666 | 5.38 |
| National forest | 6 | 290 | 0.01 | 12 | 358 | 0.02 | 18 | 648 | 0.03 |
| Reserve prohibited extraction | 6 | 894 | 0.04 | 0 | 0 | 0.00 | 6 | 894 | 0.04 |
| Sustainable development reserve | 0 | 0 | 0.00 | 1 | 588 | 0.03 | 1 | 588 | 0.03 |
| Animal protection reserve | 0 | 0 | 0.00 | 0 | 0 | 0.00 | 0 | 0 | 0.00 |
| Special wildlife zone | 4 | 35 | 0.00 | 13 | 45 | 0.00 | 17 | 80 | 0.00 |
| National property in private property | 118 | 1,048 | 0.05 | 67 | 818 | 0.04 | 185 | 1,866 | 0.09 |
| Environmental protection area | 11 | 16,464 | 0.81 | 50 | 89,126 | 4.38 | 61 | 105,590 | 5.19 |
| Total | 167 | 59,897 | 2.93% | 229 | 107,878 | 5.31% | 396 | 167,777 | 8.13% |

*Source:* Authors, based on MMA (2010).

In PP Cerrado, the factors quoted as causes of the loss of original vegetation in the Cerrado included the following: (a) the area set off as nature reserves was an extremely small portion of the overall Cerrado area; (b) illegal logging for the purpose of charcoal[3] manufacturing had increased; (c) existing logging areas had not been used as agricultural land; (d) extensive livestock production had taken place in reserves and preservation districts; and (e) surveillance systems were undeveloped. As targets to be achieved by the year 2020, the plan set a decrease of at least 40 percent in the annual deforestation area, the widespread adoption of tree-planting programs, and the construction of illegal logging surveillance systems. Particularly with regard to surveillance systems, there are incentives to apply the surveillance systems for use in the Cerrado since they are already adopted in the Amazon.[4]

The Environmental Conservation Expansion Program, which is meant to effectively manage extensive areas of privately owned land using satellite imaging technology, was launched through Presidential Directive 7029 at the end of 2009, at which time the government also introduced the Rural Environmental Registry (CAR, Cadastro Ambiental Rural).[5] In addition to making registered farmers eligible for institutional financing, this granted incentives such as vegetation restoration support without the immediate application of criminal penalties, even when vegetation had been altered within the protected area. In October 2012, the Ministry of Environment issued a decree which established the obligation of registering all agricultural land in Brazil in CAR within a year (or two years, if authorized by the President).

As of July 2013, 25 states and federal districts (DF) had agreed to participate with CAR. The National System of CAR (SiCAR, Sistema National de Cadastro Ambiental Rural) was established by Presidential Directive 7830 in October 2012. Moreover, Brazil's environment conservation policy was further strengthened by Law 12651 and Law 12727 (amendment to Law 12651), enforced in 2012, which included a new regulation regarding the width of riparian forests. Today, the latter two laws are known as 'the New Forest Code' (Novo Codigo Florestal).

---

Box 5.1 The New Forest Code and CAR

Brazil's agricultural production volume has skyrocketed since 2000. For example, total grain production doubled from 100 million tons in the 2000–2001 agricultural year to 204.5 million tons in the 2014/2015 agricultural year. This increase in agricultural production is largely due to, externally, strong demand from international markets and, domestically, an expansion in farmland boundaries and improved productivity. At the same time, the expansion in the borders of agricultural land has meant an expansion in the agricultural frontier, which has led to a decrease in the original vegetation. Moreover, higher productivity has amplified concerns over environmental damage, such as the degradation of soil and water resources as a result of the use of fertilizers and pesticides. Brazil has designated the Amazon, the world's largest tropical rainforest, as a hotspot for diversity. On the other hand, the Cerrado region offers substantial potential for agricultural development, which has drawn the attention of various international environmental conservation groups with various motives.

The New Forest Code emerged from this political environment. This is a groundbreaking document that lays out the principles for agricultural development and environmental conservation in Brazil in 14 chapters and 83 articles, divided by theme, such as Permanent Protection Áreas (APPs, Areas de Preservação Permanente), Restricted Use Areas (Áreas de Uso Restrito), Legal Reserve (Reserva Legal), and CAR. The code is groundbreaking primarily because it mandates the recovery of original vegetation in areas that were deforested in the past and also stipulates in detail the extent to which the original vegetation must be restored in riparian forests (*mata ciliar*) by the scale of the river and the scale of the farmland.

The New Forest Code also designated CAR as the instrument that would ensure its effectiveness. Brazil has a massive land area; even if a law goes into effect, monitoring is rarely thorough enough and violators cannot always be detected. However, CAR, which uses cutting-edge technology, is expected to solve this problem. CAR aims to register approximately 5.2 million farms across Brazil (total land area of 329 million ha) so that environmental conservation can be monitored on all farmland.

The Minister of the Environment, Izabella Teixeira, has stated that the Code is a sign that Brazil's environmental conservation policy is starting a new chapter, and that Brazil has one of the strictest laws on forest conservation in the world. However, the Code has been criticized by agricultural proponents for an over-emphasis on environmental conservation and an inadequate agronomic basis – for example, in the standards for riparian forestation. At the same time, environmental conservation groups criticize it for its failure to prevent further expansion in farmland boundaries.

## 5.2 Japanese–Brazilian cooperation for Cerrado agriculture and Cerrado environmental conservation

Japanese–Brazilian cooperation in the Cerrado was first considered in the early 1970s, nearly 20 years in advance of Rio 92; the first document regarding coordinated agricultural development was signed in 1974. From the outset of the joint Cerrado Development Program (PROCEDER), both Brazil and Japan espoused the belief that "[a]mong the PRODECER principles, the concept that there is no sustainable agriculture development without harmony with the environment is implicit" (Brazilian Ministry of Agriculture, Livestock and Supply and JICA, 2002, p. 36 of part 5).

As a consequence, environmental conservation measures were implemented from the outset of PRODECER. As well as strict adherence to the 20 percent reserve requirement in plantations, in order to prevent the reserve areas from becoming a haphazard patchwork of isolated areas, efforts were made to create concentrated areas where individual reserve lands were joined through a 'condominium' model, as well as to form micro-corridors of reserve land made up of individual reserves (both of which will be discussed later). Moreover, measures were promoted to preserve agricultural environments, such as the introduction of contour cropping, crop rotation, and no-till farming (direct planting).

From its early stages, Japanese–Brazilian technological cooperation for Cerrado agriculture (carried out by EMBRAPA and JICA),[6] pursued in tandem with PRODECER, also sought to emphasize environmental protection in agricultural development. Such initiatives began 15 years prior to Rio 92. Specifically, in PRODECER and Research Cooperation (1977–1992), efforts were made to minimize the impact on the environment and to establish sustainable and environmentally friendly agricultural technologies and processes that would preserve the Cerrado's natural resources.

From its earliest planning stages, CAMPO, which was responsible for implementing and coordinating PRODECER, paid proper attention to the environment. With respect to reserve areas, two forms of reserve were adopted: individual reserves and joint reserve areas. As a condominium strategy was adopted for the joint reserve areas in some of PRODECER sites, and individual reserves were lumped together in other sites, the natural vegetation was preserved in large units, which helped to protect species that required expansive habitats, thereby enabling the preservation of greater biodiversity. Furthermore, the individual reserves were usually located along water arteries that flow through the agricultural land. Dubbed 'micro-corridors,' these areas served to conserve water resources, maintain biodiversity, protect against soil erosion, preserve water quality, and allow local fauna to circulate freely. Along these water arteries, steep-sloped riverside forests and marshes were also reserved to protect areas rich in biodiversity. Current laws stipulate the width of forestation, based on the width of the river, required on both banks of a river in order to preserve natural vegetation (riparian forest, or *mata ciliar*). As such, PRODECER promoted environment and ecological preservation through condominium strategy and other initiatives related to reserves. Current law reinforced this consideration with legal framework. PRODECER's initiatives in this regards are considered pioneering efforts.

In addition, in PRODECER sites, wide-ranging technical guidance and environmental enlightenment activities were provided to farm settlers by CAMPO and agricultural cooperatives. These tutorials covered a wide range of topics, including direct planting (non-tillage seeding), soil conservation through contour cropping, the prevention of soil degradation through crop rotation, the introduction of organic substances and microbial strains of *Bradyzhizobium spp.* (nodule bacteri selected by CPAC, and so on) as an alternative to nitrogen fertilizers, the introduction of biological control technologies to deal with disease and insect control, and tree-planting along rivers, in small green areas, and deforested land.

Environmental protection measures also included the establishment of windbreak forests to prevent the loss of topsoil due to wind erosion, banning the use of combinations of pesticides, taking care when diluting pesticides to avoid the pollution of water flows, and promoting adherence to environmental conservation by training laborers employed in farming. Through these efforts, the PRODECER projects, together with cooperative research, have striven to introduce environmentally friendly agricultural technologies, and to achieve sustainable agricultural production, while addressing such issues as environmental pollution.

Efforts aimed at Cerrado environmental conservation were further strengthened after Rio 92. Two new Japanese–Brazilian cooperation plans were initiated – namely, the Cerrado Agricultural Environmental Conservation Research Project (1994–1999) and the Cerrado Environmental Monitoring Survey (1992–2000).

In the Cerrado Environmental Monitoring Survey, conducted from 1992, JICA sent experts on long-term assignments to CAMPO, dispatched a survey team of experts on short-term assignments each year, and, over the course of eight years, conducted a survey of the impact of PRODECER sites in coordination with the Cerrado Agricultural Research Center (CPAC). Soil erosion, water quality and quantity, vegetation, and insects were chosen as monitoring indicators, and fixed-point observations were made for each. The results were published in 2000 in Yoshii et al. (2000), and the associated survey data and survey methods have become valuable resources in the pursuit of sustainable agricultural development.

At the same time, Japanese and Brazilian research cooperation at CPAC identified the establishment of sustainable agricultural technologies that aim to strike a balance between agricultural production and environmental conservation as the main research focus. In 1994, the Cerrado Agricultural Environmental Conservation Research Project set four targets in technical cooperation, namely: (a) the assessment of agricultural and environmental resources using remote sensing technology; (b) improvements to soil degradation through the introduction of green manure and no-till farming; (c) integrated control of disease and insect damage (i.e., not solely through the use of pesticides); and (d) the introduction of sustainable cultivation technologies for soybeans (establishment of crop rotation systems). Each of these targets has been achieved.

Brazil's and Japan's efforts at environmental and ecological conservation have been successful; furthermore, PRODECER was never subject to criticism in terms of Cerrado environmental issues. In the future, these experiences with Cerrado agriculture will likely serve as valuable reference cases for both Brazil's and Japan's international cooperation efforts aimed at tropical sustainable agricultural development.

## 5.3   Land use of the Cerrado from an environment conservation perspective

Despite the increase of agricultural production in Cerrado in the last three to four decades, land used has not increased as fast as production. This is due to significant improvements in yield per hectare. According to the Brazilian Institute of Geography and Statistics (IBGE) farm census, 61.36 percent of the growth of agricultural production (soybeans, rice, edible beans, corn, cotton and coffee) in the Cerrado during the period between 1970 and 2006 was provided by yield growth, while the remaining 38.64 percent was due to an expansion of the planted area. The yield share of the total production growth varies depending on the period and the crops. The increases in yield were mainly achieved thanks to the technological development of EMBRAPA. *The Economist* emphasized this aspect in these terms: "the availability of farmland is in fact only a secondary reason for the extraordinary growth in Brazilian agriculture. If you want the primary reason in three words, they are EMBRAPA, EMBRAPA, EMBRAPA" (The Economist, 2010).

As a result of the development of Cerrado agriculture, the area used for crops amounted to 21,590,000 ha in 2002, occupying 10.5 percent of the total Cerrado biome, which is estimated at 204,700,000 ha, according to an analysis based on Landsat satellite images.[7] The same analysis showed that about 26 percent of the Cerrado, 54,150,000 ha, consisted of improved pasture (or cultivated pasture) in 2002. Hence, the sum of cropland and improved pasture, which could be considered as total farm land for agriculture and livestock production, amounted to 75,740,000 ha, equivalent to 37 percent of the total of Cerrado. As there are urban areas in the Cerrado, which amount to 0.4 percent of the whole, and reforested areas corresponding to 1.5 percent, the total area with some kind of land use was equivalent to 39 percent of the Cerrado's total land. The other 61 percent had undergone no changes attributable to human activity, and were considered to be the natural Cerrado. Dr. Edyson Eiji Sano, head of the Remote Sensing Center of IBAMA, found it remarkable that 61 percent of the Cerrado was still covered by natural vegetation, as the 2002 land use map showed. He stated: "[t]here is no other region in the world with such intense food production preserving such a high level of natural vegetation" (see Chapter 8).

The spatial distribution of land use in the Cerrado is shown in Figure 5.1,[8] where light gray space represents natural vegetation. Black and dark gray spaces represent cropland and improved pasture, respectively.

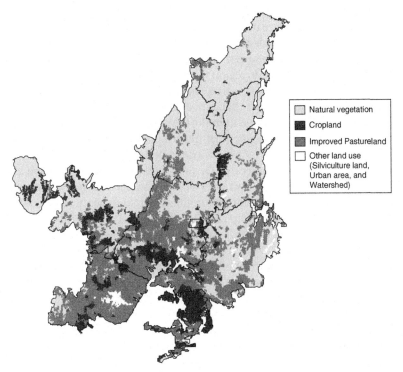

*Figure 5.1*   Spatial distribution of land uses in the Cerrado

It should be remembered that, in addition to the above-mentioned natural vegetation of the Cerrado, there are reserves equivalent to 20 percent of privately owned land. Moreover, as discussed in Section 5.1, several new initiatives have recently been introduced to protect the ecology and environment of the Cerrado. Dr. Sano mentions that the deforestation of the Cerrado has shown a decrease since 2002. The area deforested each year was reduced from 15,701 km$^2$ (annual average) during the period of 1994–2002 to 14,179 km$^2$ in 2002–2008 (0.69 percent/year), 7,637 km$^2$ in 2008–2009 (0.37), and 6,469 km$^2$ for 2009–2010 (0.32) (see Chapter 8). The total deforested area as of 2010 was 48.5 percent of the Cerrado, according to the ministry of Environment.[9]

## 5.4 Impact of Cerrado agriculture on the Amazon rainforest

During the conception and implementation of Japanese–Brazilian cooperation aimed at the development of Cerrado agriculture, the possibility that Cerrado agriculture might have an adverse effect on the ecology or environment of the Amazon region was not the major concern. On the contrary, the expectation was that the expansion of agricultural development in the barren Cerrado region would act to contain negative impacts on the Amazon, and provide what could be called a cushioning or buffering effect. It was thought that agricultural production in Brazil would expand due to the development in the Cerrado region, which would then retard agricultural development in the neighboring Amazon rainforest, representing a benefit in terms of conservation of the Amazon biome.

However, the development of Cerrado agriculture began to have an adverse, albeit indirect, effect on the Amazon for the following two reasons. First, Cerrado agricultural products started to be shipped via the Amazon River, and roads from the Cerrado region through the Amazon became wider. Second, new tropical varieties developed for Cerrado agriculture also came to be used in the Amazon.

In April 2003, Cargill drew worldwide attention when it constructed its soybean export terminal port and 60,000-ton storage silo in Santarém, located midstream along the Amazon River.. The soybeans were directly loaded onto a dedicated Panamax-class crop-carrying vessel via a 360-meter conveyor belt from the riverside silo. From there, the crop carriers could travel directly to Asian countries via the Panama Canal. Before Cargill's construction of the soybean terminal, the Maggi Group had developed an export route using the Port of Itacoatiara, the first to use the Amazon River. (See Chapter 3). However, this route greatly differed from that of Cargill. While Maggi's route passed through tributaries of the upper Amazon region from the Port of Porto Velho, Rondônia, onwards, Cargill's route, passed along a road constructed through the Amazon rainforest from the Cerrado region.

The Port of Santarém is situated at the northern end of National Route 163, and there were expectations that this road would be paved as a shipping route for Cerrado grain production. As shown in Figure 3.8, National Route 163 runs longitudinally from south to north, as if driving a wedge into the heart of the Amazon rainforest. Since there were also rumors that major crop producer Bunge would acquire land for an export base at the Port of Santarém in much the same way, there was concern that the paving of National Route 163 would go ahead and that the Amazon rainforest along the route would be destroyed. These concerns were supported by the fact that, to date, there have been strong indications of a correlation between the destruction of Amazon forests and road construction and improvement.[10]

There is a relationship between this situation and the use in the Amazon of a soybean variety developed for growth in tropical regions and widely used in the Cerrado. In Chapter 1, it was mentioned that one of the biggest factors in the success of Cerrado agricultural development was the cultivation of a tropical soybean variety. Soybean varieties less sensitive to photoperiod (the physiological reaction of organisms to the length of day or night) have been bred one after another, and coupled with improved soybean cultivation techniques; as a result, production volume in the Cerrado region has risen dramatically. At the same time, this innovation in tropical soybean varieties has meant that, technically, soybean could also be cultivated in the vast Amazon region right on the equator, where there is no variation in the length of the day or night. In fact, by the late 1990s, soybean cultivation had moved beyond the transitional vegetation zone between the Amazon and the Cerrado, and was being cultivated in the Amazon region. Soybean cultivation areas were first captured in official government statistics in 1997 for the state of Pará, and in 2001 for the state of Amazonas. This led to the Amazon rainforests being regarded as suitable locations for soybean cultivation.

Another concern was that the transnational grain majors (GTCs) would spend massive funds to stimulate soybean production in the Amazon region, which would have an overwhelming influence on global soybean production and distribution. The shipment of soybean from the Port of Santarém on the Amazon River was seen as the start of a large-scale, soybean-oriented agricultural development in the Amazon.

In these circumstances environmental protection groups, Amazon researchers, and the media collectively began to issue warnings. In September 2000, *Time* magazine published a special feature in which it asserted that the destruction of the Amazon environment was taking place due to three major factors: the development of roads, deforestation, and prescribed fires. In November of the same year, Brazil's largest weekly publication, *Veja*, also published a special feature. The magazine reported the shocking warning that by the year 2020, 42 percent of the Amazon rainforest could be destroyed, 95 percent of the total could be degraded, and only 4.9 percent of native forests could remain. Science and environmentally oriented overseas journals also ran a steady stream of articles containing similar warnings. In response to this press coverage, a boycott of Amazon soybeans was initiated in Europe, and placards along the lines of "Don't eat fast foods that devour the Amazon rainforests" (referring to broiler chicken raised on soybeans) were paraded in the streets. News of the European boycott of Amazon soybeans also received extensive media coverage inside Brazil.

Cargill's Amazon export base construction, which was taking place under these circumstances, did little more than add fuel to the fire, and resulted in heightened international concern. As a result, the ABCD group of grain majors active in Brazil,[11] the National Association of Grain Exporters (ANEC), the Brazilian Vegetable Oil Industries Association (ABIOVE), and major Brazilian companies such as the Maggi Group came to be labeled as the main culprits behind the widespread destruction of forests. Recognizing the risk of their products being boycotted by European markets, they struck a compromise with domestic and overseas NGOs[12] and explored ways to put the brakes on increasing soybean production in the Amazon's tropical rainforests.

On July 24, 2006, an agreement was reached on a groundbreaking private-sector-driven scheme known as the Soy Moratorium, whereby the parties agreed to directly inhibit "soybean production regarded as being on a forest destruction vector"[13], as a response to emergency circumstances.

The details of the moratorium comprise the following five points: (1) From July 24, 2006 onwards, companies under the jurisdiction of the National Association of Grain Exporters (ANEC) and the Brazilian Vegetable Oil Industries Association (ABIOVE) must not provide any production funds for, or make any purchases of, soybean produced on newly deforested lands or public lands (federal land, state land, reservations, etc.). (2) Whether or not soybeans have been grown on newly deforested lands is to be determined with the cooperation of the

National Institute for Space Research (INPE), using satellite images from Remote Sensing GlobalSat[14] and through visual confirmation made from the air and on the ground, and the results will be made public; the costs of this surveillance process are to be borne by ABIOVE. (3) The applicable regions are the states of MT, RO, and PA (Para), which account for 98 percent of soybean production in the Amazon region and where forest destruction had increased. (4) The Working Group will be made up of the ABCD group, ANEC, ABIOVE, major Brazilian soybean producers such as the Maggi Group, the Banco do Brasil (the government financial institution and the main source of agricultural loans), as well as domestic and overseas NGOs; the Working Group will receive all materials from GlobalSat and also take part in monitoring. (5) The initial duration of the Soy Moratorium was two years. The moratorium was effective, and has since been continuously extended. In October 2011, an agreement was reached to extend the moratorium until January 31, 2013. Later in November 2014, the moratorium was further extended to May 2016. In addition, in June 2008, when the Soy Moratorium was first up for renewal after its initial two years in effect, the Ministry of the Environment became a moratorium member.

Several technical problems have emerged with regard to the moratorium. The main issues have been: (a) the soybean cultivation season coincides with the rainy season, and the occurrence of cloud cover greatly hampers surveying; (b) the owners of designated fields are not identified; and (c) above-ground access becomes extremely poor during the rainy season. However, the public disclosure of GlobalSat reports and the fact that a national monitoring system was launched in response to widespread media coverage were major breakthroughs.

The expansion of the soybean cultivation area in the Amazon rainforest region was in fact suppressed: between the 2006 and 2011 seasons, the soybean cultivation area on newly deforested land was just 11,698 ha. It was also announced that this area represented no more than 0.39 percent of the total newly deforested area (roughly 3 million ha) in the three monitored states during the same period. Upon learning these results in October 2011, the Environment Minister declared that the soybean vector for Amazon forest destruction had been brought under control.[15] The Soy Moratorium Working Group made similar remarks. These achievements have been acknowledged by major Western consumers of Brazilian soybeans, such as Carrefour, McDonald's, and Wal-Mart.[16] The Soy Moratorium is regarded as an excellent example of improved governance of the Amazon region, made by the government, the private sector, and environmental protection NGO groups.

That said, even when areas are newly deforested, it takes a few years for the fields to support mechanized agriculture. Consequently, monitoring and regulation will also be required in the future.

Since the Soy Moratorium produced remarkable effects with regard to forests, in 2009 a similar schedule was applied to beef cattle being produced in the Amazon. After deforestation, land in the Amazon region is being used for extensively managed grazing pasture and there was considerable pressure from overseas importers and consumers for Amazon forest conservation. The numbers bear this out. Brazil is the world's largest exporter of beef, and had a herd of 208 million head of cattle in 2011. Over the previous ten years, this number increased by around 78 million. In contrast to other stagnating regions, the North region (which largely corresponds to the Amazônia Legal, or Legal Amazon region) has undergone significant growth. During the same period, the North region added 11 million head of cattle, and currently has a total of 37.9 million. Statistics on land usage (the area of regions where vegetation has been altered by human activity) in the Legal Amazon first released by Brazilian government agencies showed a cumulative 720,000 km² of land usage as of 2008, with 62 percent of this total area being used as grazing pasture; 21 percent is listed as degraded forest, and 4.9 percent is dry fields. The government has also acknowledged that cultivated pastures are the main factor in the destruction of the Amazon forests.[17]

In 2008, beef-related companies such as slaughterhouses, meat-packers, and the Brazilian Beef Exporters Association (ABIEC) reached an agreement that they would no longer purchase beef cattle or products processed from beef cattle produced on land newly deforested and converted into pasture, or on reservations in the Amazon region. Since grazing pasture accounts for the highest percentage of utilized farmland in the Amazon region,[18] this move was expected to play a large role in bringing forest destruction under control.

## 5.5  New cooperation between Japan and Brazil on environmental conservation

After PRODECER finished, a new Japanese–Brazilian project was launched with the aim of preserving ecosystems in the Amazon and Cerrado regions of Brazil. Initiatives included helping to protect the rainforest through the surveillance and prevention of illegal logging in the Amazon, as discussed previously, and protecting ecosystems in the Cerrado by reinforcing so-called ecological corridors.

Brazil is the only country in the world to have adopted a coherent surveillance system to protect its tropical forests. Since 1988, the Brazilian National Institute for Space Research (INPE) has published statistics showing the area of forestland destroyed each year, based on satellite images, and has continued to issue warnings regarding excessive deforestation. In 2004, the Plan of Action for the Prevention and Control of Deforestation in the Amazon (PPCDAM) and the Real Time Deforestation Detection System (DETER, Detecção de Desmatamentos em Tempo Real) were initiated, making it possible to monitor logging activity on a fortnightly basis. The federal police and the Brazilian Institute of Environment and Renewable Natural Resources (IBAMA), part of the Ministry of the Environment (MMA), began using satellite images provided by INPE to clamp down on illegal logging. There was, however, one major shortcoming in this surveillance system: the dense clouds that cover the Amazon during the rainy season were preventing the satellite's optical sensor from monitoring conditions on the ground. The breakthrough needed to resolve this issue came in the form of Japanese satellite technology.

In August 2006, JICA sent Dr. Manabu Kawaguchi, an expert in satellite image analysis, to Brazil for a period of six months to help IBAMA and the federal police to improve their geographic information system to prevent environmental crime. Dr. Kawaguchi concluded that the issue could be resolved by using data from DAICHI, the Advanced Land Observing Satellite (ALOS) operated by the Japan Aerospace Exploration Agency (JAXA), and he started to put his theory into practice. Launched by JAXA in 2006, ALOS uses a Phased Array type L-Band Synthetic Aperture Rader (PALSAR) microwave sensor instead of an optical sensor. This enables it to get a clear picture of the ground even in cloudy or rainy conditions, or at night. Whereas illegal loggers had previously been cutting down trees for half the year, hiding under cover during the rainy season, their activities were now exposed year-round, day and night. (JAXA's ALOS system is the only system in the world capable of providing satellite imagery around the clock.) There may have been a degree of skepticism about Japanese technology in Brazil at the outset, but ALOS images have proved to be hugely effective in combating illegal logging.

JAXA began providing ALOS data to IBAMA and the federal police in 2007, and in 2009 JICA launched the ALOS Satellite Image Utilization Project, together with IBAMA and the federal police, in order to improve ALOS imagery analysis capabilities. By using the results of this project, Brazil has greatly improved its satellite observation capabilities and its ability to crack down on illegal logging activities. As a result of ALOS technology and Brazilian efforts, the loss of Amazon rainforest area has decreased sharply.

According to a study carried out by Climate Policy Initiative (CPI),[19] the control of illegal logging by satellite monitoring in the Amazon reduced deforestation by more than 59,500 km$^2$ during the 2007–2011 period. Deforestation in the period was 41,500 km$^2$, which is 59 percent less than the level predicted before satellite surveillance (Electronic Edition of O Globo, dated September 5, 2013). The use of ALOS to help prevent illegal logging marks a significant achievment in Japan–Brazil cooperation effort. The MMA is now expediting the development of detailed satellite-based survey and surveillance systems similar to those used for the Amazon region for the Cerrado region, as well.[20] ALOS2, which boasts an analytical capacity, with PALSAR radar, ten times higher than the first ALOS, was launched by JAXA in May 2014.

The Cerrado Ecosystem Conservation Project, which began in 2003, was implemented through cooperation between JICA and IBAMA; it aimed to improve ecosystem-integrated management systems for the Paraná–Pirineus Cerrado Ecological Corridor, which extends over GO and the Federal District (Brasilia). Japan's experience in the management of national parks was crucial to the project.[21]

The Jalapão Region Ecological Corridor Project in TO, which was started in 2010 and is still ongoing, has also attracted attention. An ecological corridor refers to a continuous networked space that links wildlife habitats and enables the movement of biota. The Jalapão region is situated in a transitional zone of the Cerrado region that separates the neighboring Amazon and Caatinga (semi-arid) regions, and features rich biodiversity. The region is located in the MATOPIBA zone, which comprises the states of MA, TO, PI, and BA, and which is currently at the forefront of Cerrado development. The project aims to connect the nature reserves distributed across the four states into an ecological corridor. To achieve this, the project is aiming to strengthen the structure of the implementing agency, the Chico Mendes Institute for Biodiversity Conservation (ICMBio), which was established in 2007.

## 5.6   Towards sustainable agricultural development of the Cerrado: future challenges

Historically speaking, the development of uncultivated land was once thought to be the equivalent of economic growth. Cerrado development began in such a context, and today the Cerrado has transformed itself into one of the world's leading food production regions. There are also hopes that the Cerrado will become a source of renewable energy crops, particularly sugarcane; furthermore, the results of research into the cultivation of macaw palm, oil palm, and jatropha, which are now being tested in the CPAC, are highly promising. Global demand for food will continue to rise, along with that for bio-fuels, and the Cerrado region looks set to provide solutions. Achieving expanded agricultural production while conserving the environment and ecosystems (that is, seeking sustainable agricultural development) will, however, become an increasingly pressing need.

The role of government in advancing this concept of sustainable development is significant. In particular, it will be essential for government to do the following: (a) develop the policies and systems essential for sustainable development, and the related enforcement structures; (b) enhance zoning-based regulations and systems to monitor illegal activity; (c) build a framework for cooperation between producers, local residents, companies in the supply chain, environmental protection groups, and the government; and (d) support a solid technology transfer system through the existing state technical assistance and rural extension network (public and private) in order to provide users with agricultural technologies recently developed by research, development, and innovation (RD&I) institutions. This section discusses these duties in the light of the experiences gained from Cerrado agricultural development.

As the implementation of transport and logistics systems makes steady progress, the improvement of National Route 163 will serve as an example when conducting road improvements essential for agricultural development and the distribution of agricultural produce. The state of MT (and particularly the state's central region), Brazil's largest soybean producer, is situated inland, and producers there face the high cost of transporting their products to ports. To soybean growers in the area, the paving of National Route 163 up to the city of Santarém is a long-held desire. If National Route 163 could be paved in its entirety, it is expected that the export volume out of Santarém would increase from the current 800,000 mt to 10 million mt, and the economic effects would be significant. However, both the Cardoso administration (1995–2002) and the subsequent Lula administration (2003–2010) argued that there was a strong correlation between road improvements and deforestation, and postponed work on the roads until various related schemes could be developed. In fact, not more than 50 km of road were paved during the Lula administration, while 900 km of road remained unpaved. This is an example of the need for caution when implementing development projects that impact on environmental conservation.

The Cerrado region having been designated as a biodiversity hotspot, there will be a growing need to introduce traceability to assure the sustainability of the region. Sustainable land utilization that strikes a balance between development and environmental conservation must be quickly established having in mind the new perspective of Cerrado agriculture development. There are three important steps that need to be taken in order to achieve this: (a) to implement a land utilization plan that uses zoning to separate land suitable for agriculture from protection areas; (b) to pursue efforts to enhance illegal activity monitoring; and (c) to enforce laws using satellite surveillance technologies and other means.

The Soy Moratorium was groundbreaking. The parties currently involved are urging farms in the Amazon region to register with the Rural Environment Registry (CAR). CAR has been extended to the whole nation. By having the farmers themselves receive support from Geographical Information System (GIS) professionals, and register geographical information about the farmland they own, farmland owners can be identified from satellite data, making it easy to determine the presence of illegalities. This is not just a means by which farmers complying with the law can express a positive stance on environmental conservation – it also brings with it significant benefits, including access to low-interest financing for agricultural schemes. For this reason, the majority of farmers have already completed provisional registration and are awaiting final certification.

---

Box 5.2  Initiatives promoting environment-conscious agriculture at field level in the Cerrado region

According to the 2010 Municipal Agricultural Production (Produção agrícola municipal) survey, released by the IBGE in 2013, a survey of the agricultural product value of annual and perennial crops (total of 64 items) in the 5,490 districts nationwide found that 19 of the top 20 districts were located in the Cerrado region. This report provided evidence, if such were needed, of the extent to which agricultural development in the Cerrado region has enriched the local economy. Agricultural development generates employment and absorbs the working population from other regions where unemployment rate is high and also provides a rich source of tax revenue for the district administrative office. The district administration can then use this financial resource to conserve the natural environment.

The São Desidério district in the state of Bahia has the highest agricultural revenue in Brazil. This district is located in the MATOPIBA region, which has currently gone the furthest in promoting Cerrado agricultural development, and has a population of 30,000 people and agricultural production amounting to about US$900 million. This district's administrative office has set its own environmental conservation standards, and is known for practicing sustainable agricultural practices, such as mandating no-tilling farming and farm-visit monitoring of APPs.

The environmentally conscious agriculture practiced in the Lucas do Rio Verde district in the state of Mato Grosso, which was ranked No. 20 in this report, has attracted attention from around the world. This district has a population of 50,000 and agricultural production valued at around US$200 million, and was ranked eighth nationwide in the FIRJAN (Federação das Indústrias do Rio de Janeiro) Municipal Development Index. In the first (2006) National Environmental Awards, given for 'environmental conservation and sustainable development' every year by *Jornal do Brasil*, Brazil's most influential newspaper, the Lucas do Rio Verde district government won the award for best environmental practices in the District Environmental Administration category. This is a good example of the manner in which prosperity can strengthen environmental conservation at municipal government level.

Programs addressing environmentally conscious agriculture are also under way at farmer level. Since Cerrado agricultural development requires massive initial investment in land reclamation and farm infrastructure, farmers are very interested in sustainable, environmentally conscious agriculture. Agricultural research institutions, primarily EMBRAPA, undertake the technology development in this area. The technologies currently receiving the most attention are non-tillage cultivation, crop/cattle husbandry/forest integration systems, and plant protection using natural enemies, as explained in Chapters 7 and 8.

In addition, programs that foster the development and dissemination of technology that supports environmental conservation are moving ahead among small-scale farmers and at the community level. EMBRAPA's publication *Sustainable management of agri-biodiversity in the Cerrado biome and Caatinga biome with emphasis on rural communities* (2011), gives many examples, and efforts are under way to encourage their spread.

All this is not to say that the environmental problems of Cerrado agriculture have been entirely solved. Currently (2014), the most pressing issue is the explosive proliferation of the larva of the *Helicoverpa armigera* insect, a moth that feeds on crops. The continuous cropping of a broad agricultural area encourages the infestation of this moth, and there are now concerns not only about the agricultural damage it may cause, but also about the contamination of water resources due to the heavy use of insecticides. The reality is that, despite a rush to breed insects that are this moth's natural enemy, such efforts were not made in time. Learning process driven by administrative guidance backed by abundant financial resources and farmers' own efforts will continue to be essential to establish environmentally conscious agriculture in the Cerrado region.

The environmentally conscious conservation practices and agricultural technology developed in Brazil after many tribulations can offer many lessons that could be useful for other tropical countries, including countries in Africa.

# Notes

1. A region facing destruction despite boasting rich biodiversity. The Cerrado is one of 34 such locations around the world designated by the NPO Conservation International as a biodiversity hotspot.
2. Nevertheless, 35 percent is required in the case of the Cerrado biome located in the area of the Amazon tropical rain forest (Legal Amazon Area), as stipulated by the amendment of the Forestry Law enacted in June 2012.
3. There is a high demand for charcoal from Cerrado shrub forests, as it is required by the steel industry in the production of steel, making it highly profitable.
4. Remarks made by Environment Minister Izabella Teixeira on September 13, 2011 (from the news section of the Ministry of the Environment website).
5. A registry that uses a GIS to determine the borders of each farm, as well as the legal reserve and preservation areas in each part of owned land. Upon the request of a farm, expert contractors prepare digitized drawings of the land usage status inside each farm area. Electronic data are incorporated into Integrated Environmental Monitoring and Licensing databases operated by state government environmental agencies.
6. For details of this cooperation, see Chapters 1 and 6 of this book.
7. This is the result of research to map land use in the Cerrado at a scale of 1:250,000 using the methodology of Landsat image segmentation. The published research was supported by the Project of Conservation and Sustainable Utilization of Brazilian Biological Diversity of the Ministry of the Environment, World Bank Global Environment Facility and IBGE, among others (Sano et al., 2008).
8. As the definition of *Cerrado* differs among institutions such as the CPAC and IBGE, the figures above are not strictly comparable. For details, see Chapter 8.
9. This is cited from "A proteção do Cerrado", an article of *O Estado de sao Paulo*, (September 19, 2011) based on information provided by Ministry of Environment. See also Ministry of Environment (2014a).
10 When roads are upgraded, side roads are built at perpendicular angles on both sides of the main road, and deforestation is exacerbated. The tracks of these side roads are known as 'fish bones', as their pattern when viewed from above evokes this image. These fish bone structures can be easily found by searching the Amazon region using Google Earth. Three-quarters of the destruction of Amazon forest is said to have occurred in swaths of 50 km on either side of major roads.
11. An abbreviation referring to four transnational GTCs doing business in Brazil: ADM, Bunge, Cargill, and Dreyfus (Luis Dreyfus).
12. Greenpeace, The Nature Conservancy, the WWF, IPAM (Environmental Research Institute of the Amazon, a Brazilian NGO), and so on.
13. A term originally used to refer to vectors of insects carrying pathogens.
14. Known as GlobalSat from August 2007 to October 2009. The company name was changed to Geoambiente thereafter.
15. Electronic edition of *Valor Econômico* dated October 13, 2011.
16. These companies formed a group called the Amazon Alliance and take group-level action.
17. September 2011, Environment Minister Izabella Teixeira and Brazil Science and Technology Minister Mercadante.
18. EMBRAPA-INPE (2011).

19. The CPI is an NGO financed by Open Society Foundations, supported by George Solos. In Brazil, the Foundations' partner is the Catholic University of Rio de Janeiro.

20. Remarks by Environment Minister Izabella Teixeira on September 13, 2011 (from the news section of the Ministry of the Environment website).

21. JICA has made available on the Web video-based educational materials about the project to improve rainforest monitoring capabilities using ALOS. Visit http://jica-net.jica.go.jp/dspace/handle/10410/792 to view "Protecting the world's rainforests from space – coordination and cooperation between Japan and Brazil" (2011, 30 minutes).

## Bibliography

Altmann, Nilvo (2010) *Plantio direto no Cerrado – 25 anos acreditando no sistema.*

Barona, Elizabeth, Navin Ramankutty, Glenn Hyman, and Oliver T. Coomes (2010) 'The role of pasture and soybean in deforestation of the Brazilian Amazon.' *Environmental Research Letters* [online], 5, 024002.

IBGE (Brazilian Institute of Geography and Statistics) (2010a) 'Indicadores de Desenvolvimento Sustentável Brasil 2010' in *Informação Geográfica*, 7 (Rio de Janeiro: Estudos e Pesquisas).

IBGE (Brazilian Institute of Geography and Statistics) (2010b) *Atlas Nacional do Brasil: Milton Santos.*

IBGE (Brazilian Institute of Geography and Statistics) (2010c) *Produção Agrícola Municipal: Culturas Temporás e Permanentes 2010*, Vol. 37 (Rio de Janeiro).

Chiaretti, Daniela (2013) 'Fiscalização por satélite diminui em 59% o desmatamento na Amazônia.' *Valor Econômico*, May 8.

EMBRAPA-INPE (2011) 'Dados municipais do levantamento de informações de uso e cobertura da terra na Amazônia: TerraClass 2008,' http://www.inpe.br/cra/projetos_pesquisas/tabela_municipios.pdf, date accessed May 30, 2014.

GlobalSat (2009) 'Relatório: Segundo ano de mapeamento e monitoramento da soja no bioma Amazônia,' http://www.abiove.com.br/sustent/relatorio08/moratoria_carderno_2monitora_0409.pdf, date accessed May 30, 2014.

GlobalSat (2010) 'Soy Moratorium,' http://www.abiove.org.br/site/index.php?page=soy-moratorium&area=MTEtMy0x, date accessed May 30, 2014.

Góis Aquino, Fabiana de and Maria Cristina de Oliveira (2006) *Reserva legal no bioma cerrado: uso e preservação* (Planaltina: EMBRAPA Cerrados).

Grupo de Trabalho da Soja (2013) 'Moratória de Soja: Mapeamento e monitoramento do plantio de soja no bioma Amazônia – 6°Ano,' http://www.abiove.org.br/site/_FILES/Portugues/03022014-160810-port_final_baixa.pdf, date accessed May 30, 2014.

JICA (2006) *Relatório de avaliação final: projeto de conservação de ecossistemas do cerrado: corredor ecológico do cerrado Paraná–Pirineus* (Brasilia: JICA Brazil).

Klink, Carlos A. and Ricardo B. Machado (2005) 'A conservação do Cerrado brasileiro.' *Megadiversidade*, 1(1), pp. 147–156.

Machado, Altair Toledo, Luciano Lourenço Nass, and Cynthia Torres de Toledo Machado (2011) *Manejo sustentável da agrobiodiversidade nos biomas Cerrado e Caatinga com ênfase em comunidades rurais* (Planaltina: EMBRAPA Cerrados).

Marchetti, Delmar and Antonio Dantas Machado (eds) (1979) *Cerrado: uso e manejo: V Simpósio sobre o Cerrado: EMBRAPA-CPAC, CNPq.* (Brasilia Editerra).

Margulis, Sergio (2004) 'Causes of deforestation of the Brazilian Amazon,' World Bank, Working Paper No. 22.

Martinelli, Luiz, Rosamond Naylor, Peter M. Vitousek, and Paulo Moutinho (2010) 'Agriculture in Brazil: impacts, costs, and opportunities for a sustainable future.' *Current Opinion in Environmental Sustainability*, 2(5/6), pp. 431–438.

Martius, C.F.V. (1951) 'A fisionomia do reino vegetal no Brasil.' *Boletim Geografico*, 8(95), pp. 1294–1311.

Ministry of Agriculture, Livestock and Supply (MAPA) and JICA (2002) *Japan-Brazil Agricultural Development Cooperation Programs in the Cerrado Region of Brazil: Joint Evaluation Study: General Report* (Brasilia and Tokyo: MAPA and JICA).

MMA (Ministry of the Environment) (2002) *Avaliação e identificação de áreas e ações prioritárias para a conservação, utilização sustentável e repartição dos benefícios da biodiversidade nos biomas brasileiros* (Brasília: MMA/SBF).

MMA (Ministry of the Environment) (2006) 'Programa nacional de conservacao e uso sustentavel do bioma Cerrado: Programa cerrado sustentavel,' http://www.mma.br/estruturas/sbf/_arquivos/programa_bioma_cerrado.pdf, date accessed May 30, 2014.

MMA (Ministry of the Environment) (2010) *Plano de açao para prevenção e controle do dessOmatamento e das queimadas* (Brasília: MMA).

MMA (Ministry of the Environment) (2014a) PPCerrado: Plano de Ação para Prevenção e Control do Desmatamento e das Queimadas no Cerrado, 2a FASE (2014–2015).

MMA (Ministry of the Environment) (2014b) Cadastro Ambiental Rural (website), http://www.car.gov.br/, date accessed May 30, 2014.

National Institute for Space Research (INPE) (2014) Projeto PRODES (website), http://www.obt.inpe.br/prodes/prodes_1988_2013.htm, date accessed May 30, 2014.

Nishizawa, Toshie and Hongo Yutaka (2005) *Amazon – Hozen to Seitai* [Amazon – Conservation and Ecology] (Asakura-Shoten).

Oliveira, Paulo S. and Robert J. Marquis (2002) *The Cerrados of Brazil: Ecology and natural history of a neotropical savanna* (Columbia University Press).

Presidência da República do Brasil (2012), National Congress, Law No. 12.651, http://www.planalto.gov.br/ccivil_03/_ato2011–2014/2012/lei/l12651.htm, date accessed May 30, 2014.

Ramalho, Paulo Ernani Carvalho (2008) 'Cerrado: Ecologia e Flora,' Vol. 1 (Empresa Brasileira de Pesquisa Agropecuaria-EMBRAPA/Cerrado).

Sano, Edson Eyji, Riberto Rosa, Jorge Luís Silva Brito, and Laerte Guimarães Ferreira (2008) 'Mapeamento semidetalhado do uso da terra do bioma cerrado.' *Pesquisa Agropecuária Brasileira*, 43(1), pp. 153–156.

Solbrig, O.T. and M.D. Young (eds) (1993) *The world's savannas: Economic driving forces, ecological constraints and policy options for sustainable land use* (Taylor & Francis).

Spehar, C.R., P.I.M Souza and FAO (1999) *Sustainable cropping systems in the Brazilian Cerrados* (FAO).

The Economist (2010) 'Brazilian agriculture: the miracle of Cerrado,' August 26.

Warming, Eugenius and Mário Guimarães Ferri (1973) *Lagoa Santa / a vegetação de Cerrados Brasileiros* (Itatiaia: Da Universidade de São Paulo).

Yoshii, Kazuhiro, A.J. Camargo, and Álvaro Luiz Orioli (2000) *Monitoramento ambiental nos projetos agrícolas do Prodecer* (Planaltina: EMBRAPA Cerrados).

Zakia, Maria José and Luis Fernando Guedes Pinto (2013) *Guia para aplicação da nova Lei Florestal em propriedades rurais* (Piracicaba: Instituto de Manejo e Certificação Florestal e Agrícola).

# Part II

# Technological and Institutional Innovations That Enabled Sustainable Cerrado Agriculture

# 6

# EMBRAPA: Institutional Building and Technological Innovations Required for Cerrado Agriculture

*Eliseu Alves*

## Introduction

A solid and highly effective institution was considered to be essential to achieve the required innovations for the development of Cerrado agriculture. In this regard, the Brazilian Agricultural Research Corporation (EMBRAPA) was the key entity in the establishment of a strong innovation system for Cerrado agriculture. Therefore, this chapter first discusses the main principles that guided EMBRAPA and constituted the core of the 'EMBRAPA model' (Section 6.1). This aspect is important to show that EMBRAPA was ready for mature cooperation with the government of Japan through JICA, which will be discussed later. A summary of the EMBRAPA development process in terms of researchers and different types of expenditures follows (Section 6.2). The policy of Cerrado agriculture development is briefly explained in order to clarify the role that EMBRAPA has been expected to accomplish in this context. The government of Brazil chose to develop the Cerrado, and policies targeting this goal encompassed agriculture, infrastructure, research, universities (especially graduate programs), and rural and urban infrastructure. It was an all-inclusive regional development program (Section 6.3). The chapter goes on to discuss what kind of external cooperation would have been most effective for EMBRAPA to accomplish such a role, focusing on the most important issues that needed to be taken into account in formulating EMBRAPA's institution building plan (Section 6.4). To deepen this analysis, Section 6.5 distinguishes two types of external cooperation from an institution building perspective: a 'joining institutions' type and a specialized type. In this regard, the results of cooperation should be assessed in terms of two aspects: (a) institution building by which the recipient institution becomes prepared to face present and future challenges; and (b) new technology developed through cooperation and its dissemination to farmers. From this perspective, distinctive characteristics of the EMBRAPA–Japan cooperation are discussed (Section 6.6). Finally, concluding remarks are presented (Section 6.7).

## 6.1   The 'EMBRAPA model' and its main principles

EMBRAPA's role is to be ready with answers to society's problems in the field of agriculture. As a research institution, EMBRAPA has to answer to society quickly, because without prompt answers, investments will not be forthcoming, or will soon be discontinued. Let us look into the fundamental principles that guided EMBRAPA's development.[1]

For a research institute such as EMBRAPA to be successful, it should have an organizational model that is flexible and free from bureaucratic burdens. The organization must be free to formulate, adjust and manage its budget, to select its priorities and carry out research, and to formulate and administer its human resources policies; therefore, a public corporation model was chosen for EMBRAPA according to the laws of Brazil. EMBRAPA's geographical scope was to encompass all of the national territory, and it was to carry out research cooperation with other countries. Moreover, research requires competent researchers, and so it was necessary to provide appropriate training and to pay competitive wages. The institute's leadership needed to be at an international level. To be productive, researchers need libraries and up-to-date laboratories, both of which had to be integrated with advanced global research centers. The recruiting and promotion of researchers had to be according to merit, based on objective evaluations. It was essential to establish a research career path that challenged talent, rewarded good work, and offered the conditions for comprehensive dedication to research. In research, the scarcest resource is the researcher's time. Laboratories, libraries, and research support all have the role of multiplying the researcher's time. Bureaucracy exercises an opposite effect, and for this reason it had to be minimized. The units of research (research centers) needed to effectively explain themselves to researchers and society; they had to avoid the dispersion of efforts, and facilitate interaction among consumers, farmers, and agri-businesses.

To follow these principles, EMBRAPA established national centers, each of which took care of a limited number of previously established priorities. As a result, society, farmers, extension workers, and authorities all knew which center took care of what products, such as soybeans, corn, rice, and beans, and which center was responsible for the Cerrado and advanced biology research. This model facilitated the evaluation of successes and failures, and helped farmers to discover the technologies and knowledge they wanted, and evaluations and promotions became connected to the productive life of the researcher.

The organizational lines of authority are well defined in EMBRAPA, and its directors are free to enforce the rules, without partisan or third-party interference. Research funding is drawn from the national treasury, and treasury and budget authorities had to be aware of the economic value of research relative to other investments. To communicate with the authorities, research institutions require economists with high-level training to establish communication channels in appropriate language, and the knowledge of research impacts is crucial to show how research investments rank with alternatives. Journalists are also very important in terms of building a positive institutional image, and so they are at the same level of importance as the researchers themselves, and require training and resources to develop their work. Researchers must have access to the international research community, and the research budget should finance international travel, joint projects with international scientists, and Internet use.

Research institutions do not grow without the support and respect of politicians, party leaders of different ideological shadings, consumers, and government authorities. Legislation should encourage partnerships for mutual benefit between public and private research, free from legal risk. Finally, partisan politics and ideologies cannot dictate research priorities, research methods, and the choice of members of the board of directors.

## 6.2 Development of EMBRAPA

The above provided a condensed discussion of EMBRAPA's fundamental principles, which this section rounds out with data on the size of its labor force, expenditures, how it covers Brazil geographically, and its international agreements.

EMBRAPA's labor force includes researchers and research support personnel. There are also several types of arrangement, of a short-term nature, aimed at training students and non-students, but only if they have recently graduated. The detailed data document researchers and research support personnel: those who have had long-term contracts with EMBRAPA. Although short-term personnel are very important for

EMBRAPA's work, their numbers fluctuate a great deal from month to month, and hence the discussion here concentrates on researchers and support employees.

We can divide the period from 1973 to 2010 into three sub-periods. In the 1973–1990 sub-period EMBRAPA expanded vigorously: its labor force grew continuously to reach a peak in 1990. Subsequently, there was a reduction in support personnel for two reasons: first, the administration chose more efficient methods to handle its support personnel; and second, as a consequence of the economic crises, expenses had to be cut; in order to keep researchers, there was no other way than to reduce the number of support employees. This difficult period lasted from 1990 to 2005, which constitutes the second sub-period; recovery then followed in the third sub-period, which continues to the present. As EMBRAPA responds to Brazil's technology needs, its growth will advance at a slower pace than in the past, according to new demands.

EMBRAPA's short-term personnel are mainly students from the agrarian science universities, and part of their training must be undertaken outside the university in research institutions or in the private sector. In the course of their training they carry out important work supervised by senior researchers. In 2010, there were a total of 2,970 trainees, of whom 355 were postgraduate students. In December 2010 EMBRAPA had 9,248 regular employees, and so the short-term personnel accounted for 32.1 percent, or about one-third of its labor force (under contract).

Regarding researchers and support personnel, Figure 6.1 illustrates the sub-periods noted above: vigorous grow of the labor force from 1973 to 1990; a severe cut in support personnel between 1990 and 2005, with minor growth in the number of researchers; and a period of recovery from 2006 to 2010, at a pace appropriate for a mature institution where growth keeps pace with new demands.

In 1973, when EMBRAPA was inaugurated, the agrarian science universities graduated professionals at the Bachelor of Sciences (BSc) level only, but EMBRAPA required people with Master's degrees and Ph.D.s, and so it developed a vigorous program, in Brazil and abroad, to graduate researchers at the Master's and Ph.D. levels. The program was so successful that a large majority of EMBRAPA researchers were Ph.D.s by December 2010. The EMBRAPA graduate program now takes care of post-doctoral training, because Brazilian universities offer high-quality training at the Master's and Ph.D. level. As the market supplies high-level trained professionals to EMBRAPA, or to any other institutions that wish to hire them, there is no reason for EMBRAPA to invest in post-graduate training, except in a limited number of cases.

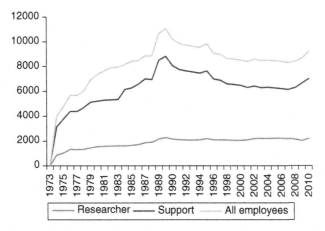

*Figure 6.1* Trends in the number of researchers, research support staff, and all employees of EMBRAPA

*Source:* Author.

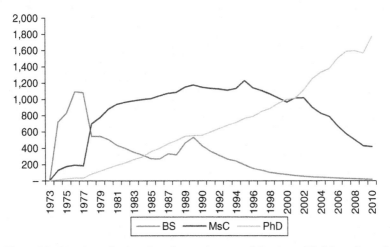

*Figure 6.2* Number of researchers by graduate qualifications: BS, Master's, and PhD

*Source:* Author.

In December 2010, EMBRAPA's labor force comprised 9,248 employees, split between 2,215 researchers (24 percent) and 7,033 research support personnel (76 percent). Of the researchers, 1,775 were Ph.D.s (80.1 percent), 421 had Master's degrees (19.0 percent), and 19 were at the BS level (0.9 percent). The graphic shows that EMBRAPA's research staff are converging toward being Ph.D. only. Looking at these numbers, one can see the three sub-periods, the success of the training program, the cut in support personnel, and the preservation of researcher positions.

Figure 6.3 shows the yearly variation in the rate of EMBRAPA expenditures as a percentage of the GDP of the agriculture sector. The GDP share line traces the behavior of the actual rate over time. The other line shows an index rate equal to 1 percent, on the assumption that total expenditures on agricultural research as a share of the agricultural GDP should equal at least 1 percent. Since the data do not cover expenditures by state research institutions, federal universities and institutes, or private sector research organizations, this record shows Brazil to be in a favorable position. There was vigorous and continuous growth in EMBRAPA's expenditures in its first decade, from its beginning to 1982, which was followed by a strong fall in 1983 and 1984 due to macro-economic adjustments. EMBRAPA's expenditures as a share of agricultural GDP then began to move up again and exceeded the 1 percent mark in 1991. In 1992 the share dipped, then rose again to reach the peak of the series in 1996. It stayed above the 1 percent mark in the period from 1995 to 2001. Again due macro-economic adjustments, EMBRAPA's expenditures as a share of agricultural GDP fell until 2003, and then moved up again to exceed 1 percent in 2009 and 2010.

The data show that the government has tended to align the growth of EMBRAPA's budget to growth in the GDP of the agriculture sector. Dips in the level of expenditures were consequences of macro-economic

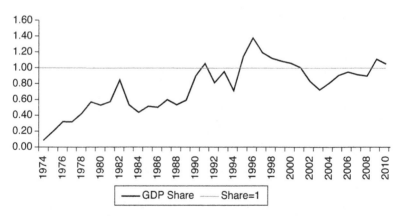

*Figure 6.3*   EMBRAPA expenditures as share of agricultural GDP
*Source:* Author.

adjustments, and EMBRAPA's budget cuts were much less severe than those of other public organizations. This is a clear indication of the importance that economic authorities in Brazil ascribe to EMBRAPA, in consideration of agricultural modernization.

The overall and item-wise trends in EMBRAPA's expenditures are shown in Figure 6.4, divided between employees (labor force; the second line from the top), capital costs (investments in capital items; the fourth line from the top), and operational expenditures (day-to-day expenditures in support of research work; the third line from the top). Overall expenditures correspond to the 'Total' line in the figure (first line). The shape of this line is quite similar to that showing expenditures as a share of agricultural GDP (Figure 6.3), which is an additional argument in favor of the thesis that Brazil's federal government is sensitive to growth in the agriculture sector in allocating resources to EMBRAPA.

As already noted, budget cuts did not touch researchers, and there were small increases in their numbers during the period of reduced budgets. This came at the expense of resources for capital and operational expenditures, reducing EMBRAPA's ability to carry out research. In some years, labor force expenditures came dangerously close to total expenditures. Recent government efforts, mainly from 2002 onward, have been devoted to overcoming this weakness in EMBRAPA's budget.

EMBRAPA is now the leader of the Ministry of Agriculture's agricultural research system, with headquarters in Brasilia, 43 national

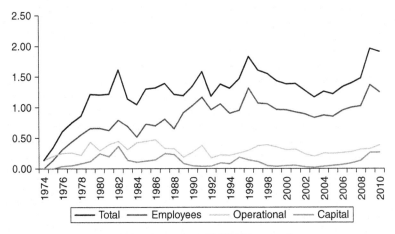

*Figure 6.4*  Trends in total and item-wise EMBRAPA expenditures
*Source:* Author.

research centers covering all areas of the nation, and 4 services units, for a total of 47 organizational units. In addition to these, there are 14 business offices. Outside Brazil, there are 5 virtual laboratories (Labex), in the United States, Europe, and Asia, and 6 foreign offices, in Ghana, Mozambique, Mali, Senegal, Venezuela, and Panama. There are 17 state organizations associated with EMBRAPA, with research and extension mandates, and there are frequent partnerships with federal, state and private universities, NGOs, and private organizations, some of them multinationals.

## 6.3   Policies to develop Cerrado agriculture and the role of EMBRAPA

The Cerrado accounts for 25 percent of the area of Brazil, and is mainly found in the Central-West region. Up to the 1960s extensive cattle ranching was the dominant activity in the context of an underdeveloped agriculture. The soil is poor and acid, and spells of drought in the rainy season are main risks facing the production of grain. The conquest of the Cerrado by modern agriculture is one of the most important achievements of agrarian sciences. It is not the objective of this section to report the complete history of modern Cerrado agriculture, but to note some facts that are relevant to Japan–Brazil agricultural research cooperation in the Cerrado.

With the inauguration of Brasilia in 1960, the federal government undertook the construction of a railroad and road grid linking the capital to the main cities of the Southern, Southeastern, Northeastern, and Northern regions. It built airports and a communications infrastructure. The federal government also created a very large program of credit to support the improvement of the Cerrado's soil and the development of modern agriculture. Population density was very low in the Cerrado, which was an advantage for farmer migration from the Southern and Southeastern regions. Compared to the local gentry, the migrants were much more advanced in terms of modern agricultural knowledge. They sold their smallholdings, bought larger areas, and settled them using modern agricultural techniques. Few among the local population 'moved up' to become modern farmers. Modern agricultural knowledge and an inclination to take risks motivated the migrants to search for advanced technologies in research centers and to take advantage of subsidized credit to support their work. The Cerrado is thus a typical case of agricultural development promoted by farmers from more advanced agronomic culture, rather than the local population, and this is a point that cannot be overlooked in dealing with Africa. In this process, the local population benefitted from technology and experiences of those newly settled farmers.

Cerrado research ranked high as a research priority of EMBRAPA, and still does so. EMBRAPA established important research centers in the Central-West region: in Brasilia, the Cerrado Agricultural Research Center (CPAC; specific to the Cerrado), the National Vegetable Research Center, the National Genetic Resource and Biotechnology Research Center, and the headquarters of EMBRAPA; outside Brasilia, the Rice and Bean National Research Center in Goiânia, and the Beef Cattle National Research Center in Campo Grande (MS). The Corn and Sorghum National Research Center is outside the Central-West region, but within the Cerrado, in Sete Lagoas (MG). The Soybean National Research Center played a very important role in developing soybean varieties adapted to low latitudes such as the Brazilian Cerrado, and without them, the development of the Cerrado soybean crop would not have occurred. New wheat varieties adapted to the Cerrado also played an important role. The national research centers for soybeans and wheat are in the Southern region, in Londrina (PA), and Passo Fundo (RS), respectively. The Dairy Cattle National Research Center, located in Juiz de Fora (MG), is also investing resources in improving milk production in the Cerrado. Moreover, the federal government established universities and Master's and Ph.D. programs in all the states in the Central-West region: Federal District, GO, TO, MG, and MS. It provided research incentives to universities and research institutes to find solutions to problems of the Cerrado, and hence a strong and diversified research network was created and developed in the Cerrado region.

Collaborations with Japan, the United States, France, the World Bank, and the Inter-American Development Bank were very important, as well as with the International Centers within the orbit of the CGIAR. The development of the Cerrado is a case of success in international cooperation: between governments, research institutions, universities, and international centers. But the Japanese cooperation was unique in the sense of institution building: of joining the researchers and resources of the two countries to develop a particular region. It was a case of cooperation where the two sides worked together with the same objectives.

In the 1970s, high grain and meat prices were instrumental in the take-off of the Cerrado program. When prices fell, the development of the Cerrado was firmly on the way, without any chance of retreating.

## 6.4 EMBRAPA's challenges and the expected role of external cooperation

This is the context in which international cooperation between EMBRAPA and multilateral and bilateral agencies was carried out. The most relevant question here is when and how external cooperation can be said to succeed, which can be answered as follows.

(1) The recipient institution becomes prepared to face the challenges of its mandate, and learns to discover upcoming problems and how to solve them.
(2) At the end of the cooperation, the recipient institution learns to stand on its own feet.
(3) The external and recipient institutions both gain by working together.

Before the start of external cooperation it is important to understand the problems of the recipient institution. The following points must be looked at in formulating an institution building plan.

(1) The bottlenecks in the way of the external cooperation: these can include a lack of counterpart funds, and problems of legal and cultural origin.
(2) The goals of the institution.
(3) The problems of the institution, its challenges, and its capacity to face them.
(4) The conceptual basis of the institution, and how compatible this conceptual basis is with its goals.
(5) The organizational model; how flexible it is, and how it adheres to the goals of the institution.
(6) The flexibility to manage the budget and to reallocate it to new demands.
(7) The flexibility to manage personnel, and identification of legal and bureaucratic restrictions.
(8) The researcher career path, the flexibility to model it, and competitive wages and compensation.
(9) In hiring, are equal rights guaranteed to those who apply for positions, and are selections based on merit? Are lack of merit and wrongdoing used to justify dismissals?
(10) The ability to communicate with society at large, and its relationship with the media and its clients.

(11) The capacity to establish research priorities.

(12) The capacity to formulate a research program, considering budget restrictions, and the authority to carry it out.

(13) The competence to evaluate itself, to estimate the returns on its investments, and to communicate its results to society, authorities, and important leaders.

(14) The capacity to convince government and the private sector to maintain and to increase its budget.

(15) The ability to relate to the world of politics based on the principles of mutual respect and independence.

(16) Hiring and promotions reward competence and loyalty to the institution.

(17) The selection of managers for research units rewards merit and managerial competence.

(18) The competence to relate to international research institutions, universities, and the private sector.

(19) The potential to retain outstanding researchers on the staff.

(20) Physical infrastructure: buildings, laboratories, experimental fields, etc.

In this regard, the concept of institution building is considered to be particularly important. We distinguish two different concepts: Institution building refers to a set of actions that are performed to change the structure of an organization, or to add to its capacity to do more of the same things in the same way. One example of the latter case is when laboratory equipment is supplied by a donor country.

In the first case, there may be changes in the organizational model, in the processes of setting priorities, planning, executing research plans, conducting evaluations, and diffusing research results. It may involve setting researcher career paths, promotions, and wages so as to reward competent work and to stimulate talented researchers. It may be necessary to strengthen the capacity of the organization to communicate with budget authorities, and with executive and legislative powers, or it may be a case of improving communications with the media at the local, regional, and national levels. The relationship with private research or with international institutions may need attention, or infrastructure may be lacking or require reforms. Which of these items receive attention depends on the nature of the problem, on the recipient institution, and on the amount of money that will be invested in the cooperation. An extreme case would be one involving an all-encompassing change by which a new institution emerges, and this was the case with EMBRAPA.[2] Such extreme cases are rare: it is more common to change selected points in the structure of an organization. One interesting case is to center the cooperation in a few research units, and to focus it within a region. The JICA–EMBRAPA cooperation focused on the Cerrado and was concentrated in a few EMBRAPA research units and state research and extension organizations.

In the second case noted above, human or financial resources are provided to enhance the capacity of an organization to do more of the same things in the same way: the structure of the organization is not changed. This is weaker type of institution building, which involves enlarging the capacity of an institution to do research, without changing its quality. In the real world, the two types are present at the same time. If the institution is mature, more weight will be placed on the second case; however, the first case is more common.

In every institution, small changes are always occurring, and they may be positive or negative in character. However, by 'institution building' we mean purposeful changes designed to improve the organization, and they are visible and sizable. Institution building may exist with or without external cooperation; for instance, the decision to transform National Agricultural Research Department (DNPEA) into EMBRAPA belonged exclusively to Brazil. After its inauguration, EMBRAPA sought technical and financial assistance from many places: international banks, other countries, research institutions, universities in Brazil and abroad, and Brazilian financial institutions, via Studies and Project Finance Organizations (FINEPs).

## 6.5   Types of external cooperation from an institution building perspective

Two types of external cooperation can be distinguished, as detailed below.

### (a)  Joining institutions

The cooperation agency (country, institution, etc.) sends staff that join the staff of the recipient organization under its rules of administration. Specialized coordination may exist to facilitate communication between foreigners and local nationals, and among the foreigners themselves. From that point on, all activities are jointly planned and carried out: as if there were an informal merging of the agencies into a unique institution for the purpose of the objectives of the cooperation project. The results of the joint work belong to both parties: it makes no sense to try to separate them. There are many spillover effects from one party to the other, which work against splitting the outcomes of the project.

Cooperation encompasses external and internal funds: the counterpart funds. In general, the research program is carried out in the recipient country; if convenient, part of it can be performed in the donor country, or even in third countries. The training program may be carried out in the recipient or in the donor country; the latter applies, in general, to graduate training. For the acquisition of laboratory and other types of equipment, the donor part of the budget may be imported, and part may be acquired locally. The cooperation program may concentrate on research, or it may move up a step further to take care of technology transfers or even investments that facilitate the adoption of technology by farmers, such as electricity and irrigation. If the cooperation succeeds, its benefits extend far beyond the end of the agreement. There are also spillover effects into regions not included in the original agreement. A careful evaluation of the project requires looking into its results at the end of the agreement and also far ahead in time, in the defined region and in other regions that may have partaken of the benefits. There are two types of result: institutional development, by which the recipient institution becomes prepared to face present and future challenges; and research results. They are designed to solve farmers' problems and to modernize agriculture. Therefore, the evaluation of the project cannot overlook institutional development, the new technology it helped to develop, and the adoption of such technology by farmers. The JICA–EMBRAPA project exemplifies the joint type of cooperation, from which EMBRAPA, the Cerrado, and Brazil all profited.[3]

(b) **Specialized cooperation**

Most external cooperation is specialized in themes, and does not join, in an informal way, two institutions for the purpose of carrying out the terms of an agreement. It may be a case in which a foreign university opens its doors to receive graduate students, and the program expenses may be paid by a grant from the donor country, by the recipient country, or by a loan from international banks. At its own cost, the donor country may send scientists to deal with specific problems of the recipient institution, and the technical assistance that follows may be of a short- or long-term nature. There may be an agreement between two countries according to which the developing country can post a group of senior scientists to the developed country to work with counterpart scientists, under the umbrella of specific organizations. This is the case of the Brazilian Labex experience, which can address research problems of mutual interest to the two countries. However, an agreement may not directly involve monetary expenditures: for instance, when two institutions, one in the developing country and the other in the developed country, open their doors to allow the exchange of experience among scientists via the Internet and other means of communication.

External cooperation of both types has been very important to the development of EMBRAPA and in particular to the development of research for the Cerrado. Of particular significance have been the Japanese, American, and French cooperation, that of the International Centers revolving around CGIAR, and that of international banks, the World Bank and IDB. Except for the Japanese experience, all of these were of the specialized type, in the sense that they emphasized themes and did not join with EMBRAPA in Brazil to solve a set of research problems.

## 6.6   Distinctive characteristics of the EMBRAPA–Japan cooperation and its results

Up to 1970 the expansion of agriculture took place in regions where the existing knowledge of agricultural practice could be applied, if necessary with few modifications, but at the beginning of the 1970s most of that type of land had been explored. The Cerrado and the Amazon region were the choices for expanding agriculture, and traditional knowledge and technology would have failed there. Hence, at that time, the expansion of agriculture required investments in science and technology: facts well understood by the Brazilian government.

At the same time, the supply of food lagged behind the growth of demand, and consequently a severe food crisis emerged in the cities. Furthermore, the country needed to increase its exports to pay international debts brought about by industrialization polices. To increase food exports, for which the international prospects were excellent, the country needed to increase its surplus over domestic consumption. Thus, domestic consumption and food for export pointed to the importance of expanding agricultural production, and traditional technology was powerless to reach that goal. Importing the technology of the advanced countries would have failed, too, because technology is locally specific. The solution was to invest heavily in the generation of knowledge, which resulted in the creation of EMBRAPA, inaugurated on April 26, 1973 (Alves, 2012).

Soon after the inauguration of EMBRAPA, its leadership understood the need to shorten the road to results, because of the strong pressure from society and the government. No other way existed than to use the knowledge of agrarian science accumulated in the developed countries to generate technology useful for Brazilian agriculture. Fortunately, EMBRAPA found the doors open for this type of cooperation. Developed countries, their universities, research institutes, and the international banks understood the importance of developing Brazilian agriculture and signed cooperation agreements. In the case of the Cerrado, however, EMBRAPA saw the need for a different type of cooperation that would go far beyond the specialized type of institution building into the informal merging of two institutions, one of which would be EMBRAPA and the other a foreign institution that would jointly work in Brazil to develop technology for the region. The government and EMBRAPA understood that the best choice was Japanese cooperation; this cooperation was requested by the Brazilian government soon after EMBRAPA's inauguration in 1973, and the technical cooperation started in 1977. The reasons for choosing Japanese cooperation were as follows.

(a) Technical assistance should consider research, extension, and financial resources to develop the Cerrado. It should extend beyond the development of research, and institutional building must incorporate those activities, and JICA was willing to do this.
(b) Japanese research was very strong in agrarian science: in land saving technology, in irrigation, and in the development of small-scale equipment and machinery for the mechanization of agriculture.
(c) Japan was a major importer of food and its imports were rising.
(d) Japan wanted to diversify the sources of its food supply, and Brazil offered good prospects.
(e) The Japanese–Brazilian population (including their descendants) was large enough to facilitate the lives of Japanese researchers and professionals in Brazil.
(f) JICA's model of work and of technical assistance was flexible to the point of meeting EMBRAPA's expectations of joint work.
(g) Agrarian sciences were well developed in some Japanese universities, which was an important point for the EMBRAPA training program.
(h) Japanese industry was very advanced in producing top research equipment. The government of Japan agreed to provide funds for the acquisition of research equipment. They allowed researchers to ask new questions and answered them without loss of time, therefore saving EMBRAPA's scarcest resource: researcher time.

(i)   EMBRAPA was a mature institution, well integrated in the development policies of the Brazilian government, and conscious of its role
to develop Brazilian agriculture, of its weaknesses, and of the need
for technical assistance.

(j)   Finally, JICA expressed its willingness and demonstrated its readi-
ness to sign an agreement to experiment in an all-inclusive program
of technical assistance to develop the Cerrado.

The Brazil–Japan cooperation embraced research, extension, rural credit,
and investments in electrification and irrigation. The details will not be
covered here, since there is an excellent report that explains the concept
of the project and how it was carried out, and evaluates the results.[4]

The strongest points of Japanese cooperation from the standpoint
of institution building were: the group of outstanding researchers and
dedicated assistants that came to EMBRAPA, and formed a harmonious
group with EMBRAPA researchers to carry out the research program; the
training program in Japanese universities; the top-quality laboratory
equipment that saved much researcher time and extended the questions
that could not be previously answered; the support for extension work,
research on the environment, soils, plants, animal diseases and insect
control, and plant breeding; the work with research and extension
organizations in MG, BA, MT and GO; and rural credit, electrification
and irrigation financing. Most of all, thanks to the Japanese cooperation, EMBRAPA research units greatly improved their ability to answer
the problems of the Cerrado.

The Cerrado Agricultural Research Center (CPAC, Centro de Pesquisa
Agropecuária dos Cerrados) owes much of what it has achieved to the
Japanese cooperation. The cooperation was also important for the devel-
opment of other researchers units, such as the Soybean Center (Centro
Nacional de Pesquisa de Soja), Vegetable Center (Centro Nacional de
Pesquisas de Hortaliças), and Cenargen (Centro Nacional de Recursos
Genéticos e Biotecnologia).

Clearly, not all the success of the project can be attributed to the
Japanese cooperation, but the joint project was fundamental in devel-
oping EMBRAPA's research capacity, and the technology generated by
EMBRAPA paved the way for the growth of production in the Cerrado.[5]

## 6.7   Concluding remarks

Japanese cooperation has been very useful to EMBRAPA development. It
helped to build Cerrado research, and the research results contributed

to the development of agriculture in the region. In terms of the total Cerrado share of the national population and land area, Cerrado agriculture has performed very well. Since the data indicate a pattern of development largely based on productivity improvement, it is clear that new technology was the main driving force of development. Moreover, the increase in yields saved land. In this sense, the development of the Cerrado has been friendly to the environment (Martha et al., 2012). The full impact of Japanese cooperation had much to do with the CPAC. Under the joint leadership of CPAC–JICA, the cooperation extended to other EMBRAPA units and state research organizations. It succeeded in generating the technology needed to modernize agriculture and is an excellent example of institution building, by which two countries join human and financial resources, focused on one research unit and on a specific region – the Cerrado – under a unified command. Japanese cooperation was all-inclusive: it dealt with EMBRAPA institution building; it joined Brazil in creating the implementation body CAMPO, and this company was instrumental in extension work and in settlement projects. JICA financed electrification and irrigation programs. The fact that the program encompassed all the activities required to achieve the development goals – research, extension, implementation of settlement projects, electrification, and irrigation financing – explains its successes.

One important point attracts attention: the two countries were ready to cooperate; Brazil needed to expand its agricultural output with the minimum impact on the area under cultivation, and Japan wanted to diversify its imports. The Japanese style of cooperation and expertise recognized the importance of institution building. As a consequence of the joint work, EMBRAPA is now competent to face new challenges; it has learned more in order to solve problems by itself; and both JICA and EMBRAPA have experienced immense gains.

## Notes

1. This section is based on Alves (2012).
2. Up to 1973, there was a research institution in the Ministry of Agriculture (now MAPA): the National Agricultural Research Department (DNPEA). The government decided to subject it to an all-encompassing reform program, from the organizational model to planning and executing research and technology diffusion. This reform created EMBRAPA.
3. In this spirit, at beginning of the 1960s, Purdue and the Federal University of Viçosa signed an agreement to develop a graduate program, which was financed by USAID and the Ford Foundation, and was extended to other universities.

4. MAPA and JICA (2002) *Japan–Brazil Agricultural Development Cooperation Programs in the Cerrado Region of Brazil: Joint Evaluation Study* (Brasília and Tokyo: MAPA and JICA).
5. What the author can do beyond a report from the vantage point of 2011 is to confirm the real success of the cooperation in institution building and to add indirect evidence about the growth of Cerrado agriculture, which exceeds that of Brazil as a whole in the case of grains, tomatoes, and beef and dairy cattle, and which has had good performance in carrots and potatoes, as representatives of vegetable crops.

## References

Alves, Eliseu (2012) EMBRAPA: 'A successful case of institutional innovation' in G.B. Martha Jr. and J.B. de Souza Ferreira Filho (eds), *Brazilian Agriculture: Development and Changes* (Brasília: EMBRAPA), pp. 143–160. (This article is a revised version of an article of the same title published in *Revista de Política Agrícola*, year XIX, special edition, July 2010, pp. 64–72.)

Martha, G.B., Jr., E. Alves, and E. Contini (2012) 'Land-saving approaches and beef production growth in Brazil.' *Agricultural Systems*, 110, pp. 173–177.

# 7
# CPAC: A Model for an R&D System for Tropical Agriculture

*Elmar Wagner, Wenceslau J. Goedert, and Carlos Magno Campos da Rocha*

## Introduction

This chapter discusses the technical cooperation developed between Japan and Brazil to strengthen the Cerrado Agricultural Research Center's (CPAC) research and development, and the EMBRAPA Cooperation System for the development of the Brazilian Cerrado. It addresses both changing demands and challenges for the Cerrado and those for the tropical savannahs of other countries. The discussion relates to the first two decades of activities by the CPAC, as defined in national projects and the program of the Brazilian Agricultural Research Corporation (EMBRAPA).[1]

The methodology adopted comprised a preliminary survey of technical-scientific work developed during the cooperation period, in addition to interviews with leaders and researchers at the time, analysis of EMBRAPA documentation regarding the creation of the CPAC, and documents from the Brazilian state research systems operating and/or located in the Cerrado region. In addition, documentation on the data processing and diagnostic analyzes made at the time (1975/1976) were consulted, along with mid-term evaluations, characterization of advances and contributions of Japanese–Brazilian research and cooperation, with the new and current state-of-the-art vision in 2010, and the projections of the CPAC's Strategic Plan of 2008–2023.

Section 7.1 discusses the CPAC's constitution, historical background, and activities. Section 7.2, the main body of the chapter, describes the important cooperation phase and the post-cooperation phase, as well as the technological innovation associated with the contribution of six teams of Japanese researchers during the period from 1978 to 1994 and the period from 1995 to 2010.. Section 7.3 looks at Japanese cooperation and its function in the three relevant research segments, related to the donation of equipment and training. Section 7.4 assesses technology diffusion activities and rural extension, with emphasis on the CAMPO and the Technical Assistance and Rural Extension Corporation (EMATER). The last Section, 7.5, includes a review and analysis of previous proposals regarding the transfer of knowledge and technologies to countries with similar tropical savannahs, particularly Mozambique, in accordance with proposals expressed in international events and the terms and conditions already in place between Japan, Brazil, and Mozambique for technical cooperation in Mozambique's Northern region.

## 7.1    CPAC: historical background and activities

The history of the research on Brazilian Cerrado can be divided into three periods: (i) from 1824 to 1967, a period mainly characterized by floristic, phytogeographic, and ecological studies, mainly within the scope of botany; (ii) from 1968 to 1974, the period preceding EMBRAPA's decision to establish the CPAC in 1975; and (iii) from 1975 onwards, with the registration of the natural patrimony, mainly in technical reports of the Cerrado Series: Informative Summaries, with 1,965 technical-scientific papers published, of which 571 (Volume 3) were published between 1967 to 1980; and in technical flash reports as well as specialized magazines.

Some geopolitical decisions affected the history of the development of the Cerrado. Emperor Dom Pedro II of Brazil made the first major decision when, in 1892, he issued a decree transferring the country's capital to the countryside of Goiás State, near the city of Planaltina, a satellite city of what is today's Federal District (DF). This decision was based on the findings of a scientific mission lead by Belgian researcher and engineer Luiz Crulz. The second decision, key to the development of the Cerrado, was made during the administration of President Juscelino Kubitschek de Oliveira (1955–1960), on the transfer of the country's capital from Rio de Janeiro to the DF, and the construction of Brasília, which was inaugurated on April 21, 1960. The third major political decision was made by the Minister of Agriculture, Fernando Cirne Lima, who invited the agronomist Otto Lira Schraeder, from the National Department of Agricultural Research and Experimentation (DNPEA), and economist José Irineu Cabral, from the Inter-American Institute of Cooperation for Agriculture (IICA), to review the national agricultural research system, which resulted in the 'creation' of EMBRAPA, having José Irineu Cabral as its first President (1973–1979).

The fourth important historical decision was made during the administration of President Ernesto Geisel (the administration included Alysson Paulinelli as Minister of Agriculture), with the creation of the Cerrado Development Program (POLOCENTRO), with the aim of strengthening research, promoting rural extension, generating agricultural financing and credit, and upgrading the road infrastructure, electricity supply, and storage facilities.

## (1) 1824–1967

Goodland (1979), with regard to the history of work conducted in the Cerrado until 1968, highlights certain contributions, such as those of K.F.P. von Martius, in the publication of *Flora Brasiliensis* (1840–1906), and in the series of narrations of missions and travel reports (Spix and Martius, 1824), as well as the work of Moore (1895), Pilger (1901), Sampaio (1916) and Hoehne (1923), regarding the Cerrado of Mato Grosso (MT); and the work of Glaziou (1905), with respect to the Cerrado of Minas Gerais (MG); and the works of Loefgren and Edwail (1897), Usteri (1911), and Navarro de Andrade et al. (1916) regarding the Cerrado of São Paulo (SP). Additionally, Huber (1974) highlighted some contributions in his review of the literature from 1850 to 1972.

Some pioneer work, such as the ecological study of *Warming* (1892, 1908); the experimental work of Rawitscher et al. (1943), which shows that water was not as much a limiting factor to the Cerrado's plants as previously thought; Schubart (1959), with the results of a 20-year research project regarding the water levels in wells; and Alvim and Araújo (1952), who analyzed the soils, relating them to vegetation. Other highlights are the 'Classic Work of Ecology', published by Berd (1953) and Alvim (1954), which highlighted mineral deficiency as the main cause of the disappearance of Cerrado plants, fire as the secondary cause, and drainage as a less important factor.

As a result of such studies, the theory developed by Arens (1958) and Ferri (1962) of 'oligotrophic scleromorfism' was ultimately accepted. Arens' theory stated that if the Cerrado's plants obtained abundant light, air, and water, they would be capable of photosynthesizing excess carbohydrates and fat. However, due to the lack of minerals in the soils, vegetation in the Cerrado had difficulty producing proteins, and consequently developing.

In addition, M.T. da Costa (1964) and J.J.R. Branco (1964) contributed to the knowledge of geology, and J.M. de Oliveira Filho (1964), V. Znamenskiy (1964), R.J. Guazzelli (1964), J. Bennema (1964), and others to the knowledge of Cerrado soils. Before the creation of the CPAC, according to Robert Goodland (1979), the only works covering the evolutionary stages of the Cerrado were those of H.P. Veloso (1946, 1948, and 1963). There are a few papers on the phytogeography and floristics of the Cerrado, such as *Phytogeography of Brazil* (1934), published by A.J. Sampaio. A paper showing a Cerrado vegetation map produced by Luiz G. Azevedo (1959) represented a major advance. The main types of Brazilian vegetation were compiled and classified by Aubréville in 1961. In 1963, Rizzini published a monograph about the Cerrado's floristics.

### (2) 1968–1974

The second period included the literature review of Mário Guimarães Ferri with the theme: one decade (1968–1977) of work developed in the Cerrado, from the chapter of the book: *Cerrado Ecology* (pp. 23–42). Ferri's citations, from 1975 to 1977, which are relevant for the purpose of this article, were allocated in the section that deals with the technological innovation created by the CPAC. The bibliographic citations of this eight-year period (1968–1974) can be grouped as follows: (i) botany and ecology; (ii) soils and water; and (iii) symposia on the Cerrado: I Symposium on the Cerrado: Ecological Studies (1963), II Symposium on the Cerrado (1966); and III Symposium on the Cerrado, held in São Paulo in 1971.

Analyzing studies and research on bibliographical production during these two periods, thematically and by the number of citations or references, it appears that the first period of nearly 150 years concentrated on the Cerrado flora, on issues concerning vegetation and related matters; the second period of eight years was characterized by the first appearance of other themes essential to the knowledge of natural resources, such as the first assays on soils and water, and an investigation into themes concerning climate. Such issues were highlighted upon the completion of the 'Symposium on the Cerrado' series, notably the third of the series, which was still within the scope of the University of São Paulo (USP).[2]

Brazilian and associated researchers had to overcome barriers and challenges marked by ignorance and skepticism, nearly relapsing into the concept that Cerrado soils were arid, toxic, and unfertile, and were inadequate for the production of food, beverages, fibers, essential oils, and pharmaceuticals, among other products of animal and vegetal origin.[3]

### (3) 1975–2010: foundation and activities of the CPAC

The third period, in which the authors of this article actively took part, also records important events in addition to the Symposium on the Cerrado series, such as: congresses, conferences, field trips, publication of articles in scientific magazines, and information and/or communication in general. The results of this period, under the coordination of the CPAC and its collaborators, were achieved as part of the technological innovation process of the center's research team.

It is important to understand the Brazilian time period between 1948, when a rural extension model was pursued, and 1972, when it was realized that Brazil did not have enough technical knowledge to implement the transfer and technical assistance/extension rural model. With this, the Ministry of Agriculture decided to review the national agricultural research and experimentation system.

In 1948, Brazil requested support from the government of the United States of America to redefine Brazil's rural extension model. The first experiment was carried out in MG, in 1954, with the Credit and Rural Assistance Association (ACAR-MG), an institutional figure of the third sector, adopted for all of Brazil's states and the Federal District, with the exception of São Paulo, which had created the Houses of Integrated Technical Assistance (CATI), still in operation to this day. The combination of ACAR's divisions created the Brazilian Credit and Rural Assistance Association (ABCAR). This system operated in close cooperation with Rural Social Services (SSR) for the benefit of families and rural production.

In 1973, EMBRAPA was created as a result of an agricultural research system review, replacing the DNPEA. EMBRAPA was initially organized into regional research centers, and then into 14 centers for priority products, following the model of the International Centers of the Council Group for International Agriculture Research (CGIAR). After initiating the implementation of these product centers, governmental development programs of the Mid-West (POLOCENTRO), the Amazon (POLOAMAZÔNIA), and the Northeast (POLONORDESTE) asked EMBRAPA to establish the Agricultural Research Center of the Cerrado (CPAC) in Planaltina, DF; the Agricultural Research Center of the Humid Tropics (CPATU), in Belém, PA; and the Agricultural Research Center of the Semi-arid Tropics (CPATSA), in Petrolina, PE. The National Center of Genetic Resources (CENARGEN) was created in Brasília, later aggregating biotechnology. Currently, EMBRAPA has 47 research units distributed throughout the entire national territory, and 5 virtual laboratories abroad.

In 1975, the institutional model of the Third Sector, the ABCAR system, was transformed into the Brazilian Corporation of Technical Assistance and Rural Extension (EMBRATER), which was shortly thereafter made obsolete as a public administration body. Its functions were then performed for a short period by EMBRAPA, and then moved under the direct administration of the Ministry of Agriculture.

With regard to POLOCENTRO's research component, aiming to generate knowledge for the development of the Cerrado, a work group instituted by Resolution No. 040/1974, according to the institutional model for the implementation of agricultural research at EMBRAPA, was assigned to prepare a Pre-project of Implementation of the Research Center for the Development of Cerrado Resources. The strategic location of the new Brazilian capital, the core area of the Cerrado, and its representation led to the selection of the Federal District, in the administrative region of Planaltina, where the former Agricultural Research Center of Brasília (CPAB) had operated.

A series of studies mentioned in the Pre-project of Implementation of the Research Center for the Development of the Cerrado contributed to characterize the economic and social importance of the Cerrado, as well as the decision to prioritize agricultural research and development. These included the works of Alfredo Lopes (1974); Freitas et al. (1960, 1971, 1972); MacClung et al. (1958, 1961); Mikkelsen et al. (1963); Sanchez et al. (1974), G. Eiten's *The Cerrado Vegetation of Brazil* (1972), Goodland's Ph.D. dissertation "An Ecological Study of the Cerrado Vegetation of South Central Brazil", published in Canada in 1969; the Institute of Applied Economic Research (IPEA)'s *Current and Potential Utilization of the Cerrado* (1973); and the three Symposiaon the Cerrado – 1963, 1966 and 1971.

The decision to establish the CPAC (Deliberation No. 003/1975) resulted in the Project for the Implementation of the Center, whose main objective was to devise the Research Program, the functional organization of the CPAC, and the resources necessary for its implementation.

The initial research program continued with cattle husbandry, considering the origin and program of CPAB, but soon after, upon the formation of an adequate technical team, it devised three regional projects: (i) Inventory of Natural and Socio-economic of the Cerrado's Resources; (ii) Utilization of Soil-Climate-Plant of the Cerrado's Resources; and (iii) Development of New Production Systems and Improvement of those Currently Used in the Cerrado. These three projects were to address the priority problems described in the general analysis of limitations hindering the agricultural development of the Cerrado:

1) Size (20–25 percent of the national territory) and heterogeneity of the area;

2) Insufficient knowledge of natural and socio-economic resources;

3) Low natural soil fertility;
4) Insufficient water;
5) Lack of knowledge regarding the capacity of cultivated plants to adapt to the Cerrado's conditions;
6) Occurrence of diseases, pests, and weeds;
7) Lack of agricultural production systems knowledge; and
8) Lack of infrastructure.

Initially, the CPAC's research program sought to meet the guidelines of the POLOCENTRO program, particularly those of the 13 poles with occurrences of limestone mines, and the Special Program of the Geo-economic Region of Brasília (POLOBRASÍLIA). However, at the time, the center's mission was defined as the promotion of fundamental research necessary for the generation of technologies. Thus, in addition to its initial program, the CPAC was charged with conducting basic research, in addition to contracting other institutions, EMBRAPA Products Centers, and, especially, universities. Since its beginning, the CPAC considered the ecological balance of the Cerrado region, pursuant to the sustainable development concepts established at Stockholm 1972 and Rio 1992).[4]

The eight limiting factors mentioned above became six major problems for research, referred to by CPAC researchers and technicians as the 'Big Six.' The dimensions and heterogeneity of the Cerrado and its lack of infrastructure were in fact not problems for agricultural research.

The CPAC began to operate on July 1, 1975, with efforts made for its implementation and the structuring of its multidisciplinary team, especially oriented to the Production Systems Project. In 1976, the Project of Natural Resources Utilization was strengthened in terms of its technical team, and pursued to establish the basic research principles in systems through central and satellite experiments. In 1977, the Project of Natural and Socio-economic Resources Inventory was emphasized, promoting regional coordination, and then starting technology diffusion (see CPAC 1976/77 Program; Figure 7.1).

The FAO/ANDA/ABCAR Project consisted of more than 3,000 tests, in different locations within the Cerrado and throughout the rest of Brazil. Results were obtained in terms of pedology and soils and reported in CPAC's annual technical reports, the annual symposia on the Cerrado, and the *Brazilian Agricultural Research* magazine (PAB), among other media.

From the beginning of the 1980s, the EMBRAPA program was reorganized and the CPAC became responsible for three National Research Programs (PNPs): (i) Inventory of Natural and Socio-economic of the

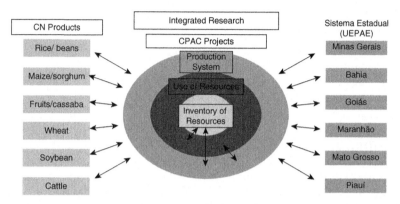

*Figure 7.1*   Research Projects' Integrated Activities – direct and coordinated execution – of the Research Center along with EMBRAPA's Products Centers
*Source:* Authors.

Cerrado's Resources (PNP Resources Evaluation), with 10 projects; (ii) Utilization of Soil-Climate-Plant of the Cerrado's Resources (PNP Utilization), with 5 sub-programs (soil fertility, soil biology, soil management and conservation, water deficiency, mechanics and energy supply) and 12 projects; and (iii) Development of New Production Systems and Improvement of those Currently Used in the Cerrado (PNP Production System), with 6 sub-programs (annual crops, wheat, perennial crops, forages and pastures, plant health, and agricultural properties systems) and 15 projects. In addition to these 37 projects, 16 projects conducted by the CPAC are part of other PNPs. These include: soybean; wheat; corn; sorghum; citrus; mango; cassava; forest, beef cattle, genetic resources, and agricultural diversification.

Table 7.1 shows the evolution of the team of researchers and technicians working in the field and the important contributions from researchers of collaborating entities.

At the time of its creation and establishment, the area intended for research and experimentation at the Research Center was 540 hectares. In 1977, it became 2,140 ha, and from 1984 on it occupied an area of 3,540 ha, belonging to the MMA.

In short, 1975 represented a milestone for the Mid-West region and for other regions, with respect to: (i) the research component initiated by CPAC's establishment; (ii) the development component, caused by POLOCENTRO's requests; and (iii) the technical assistance and rural extension component resulting from the establishment of EMBRATER.

*Table 7.1*  Evolution of the CPAC's technical team and of cooperation agreements

| CPAC Team | CPAB | 1975/76 | 1977 | 1980 | 1984 | 1994 | 2004 | 2010 |
|---|---|---|---|---|---|---|---|---|
| Researchers | 14 | 42 | 53 | 80 | 101 | 98 | 96 | 93 |
| Technology Transfer | – | – | – | – | – | 5 | – | – |
| Technical Support | – | – | – | – | – | 17 | | |
| CIAT | – | – | – | – | 4 | 7 | – | – |
| JICA | – | – | – | 8 | 7 | 16 | – | – |
| CNPq | – | – | – | 5 | – | – | – | – |
| IICA/BIRD | – | – | 2 | 6 | 11 | – | – | – |
| TROPSOILS | – | – | 7 | 4 | 2 | – | – | – |
| ORSTON | – | – | – | 2 | 1 | 1 | – | – |
| CIRAD | – | – | – | – | – | 2 | – | – |
| Total | 14 | 42 | 62 | 97 | 126 | 146 | | |

*Source:* Data from the CPAC and JICA–CPAC Project annual reports.

*Figure 7.2*  Aerial view of the CPAC headquarters
*Source:* Authors.

The states of Brazil and the Federal District continue to have the active participation of EMATER. The CPAC is now called Embrapa *Cerrado* and POLOCENTRO was replaced by the Cerrado Development Program (PRODECER) in its I, II, and III phases.

## 7.2   Technological innovation generated by the CPAC

### (1) Challenges and achievements

Technological innovation can be understood from varying perspectives including management, methodological procedures, primary production, processing, food and fiber production, packaging, and tracking. In all of these fields, the Center was the research establishment that contributed the most to the modeling of the EMBRAPA system. The Cerrado, with all the difficulties related to cultivation and production, demanded knowledge generation and technological innovations in various fields. At a time when little or nothing was known, small gains were visible and noticeable. To move from disbelief to fast-tracked accreditation, no more than two to three years of positive research results were necessary.

In this pristine land, it was possible to practice most of what the new research model, from the Ministry of Agriculture, had recommended. The systemic focus, people training and qualification, technologies and knowledge diffusion, direct action and coordinating regional action, relevant project themes, international cooperation and rural extension, along with research and experimentation, aided EMBRAPA considerably to validate and consolidate the Brazilian Model of Agricultural Research.

In the area of research management and its funding, the CPAC not only followed the recommended standard, but also surpassed it by instituting the CPAC Advisory Committee, and, internally, the Technical Committee, where main decisions were collectively discussed and approved. It was equally remarkable in publications and events, and in the disclosure/diffusion of technologies.

For EMBRAPA as a whole, the CPAC was the repository of competent administrative professionals, especially in the fields of planning, resource materials, human resources, and information/communication. This kind of theoretical-practical education for on-the-job training and formal qualification is observed to date in the whole EMBRAPA network.

The generation of new knowledge and the validation of knowledge available at the time of the CPAC's creation were separately evaluated with regard to the direct action of research conducted by the Center, and by collaborating entities, which included the Cooperative Research System, coordinated by EMBRAPA. This evaluation was due to the fact that the agricultural research and experimentation projects developed were, on the one hand, few in number and expressed very little; that is, scarce and sparse, lacking evidence obtained through more elaborate experimental techniques, while on the other hand, the levels of methodological development of collaborating institutions were unequal and did not allow combined and correlated analyzes. Similarly, although within the same research program, the contributions resulting from technical and/or scientific cooperation projects received their own treatment with particular attention given to JICA–CPAC cooperation, and to entities within the Cerrado or to those working to benefit Cerrado development.

In addition to the informative material produced to justify the establishment of the CPAC, three events were fundamentally important: (i) the IV Symposium on the Cerrado, held in Brasília under the coordination of the CPAC, in 1976; (ii) the publication of the book *Cerrado: Analytical bibliography*' by Lemos (1976), with 766 references; and (iii) the publication of *Cerrado: Informative Abstracts*, with 571 quotations (EMBRAPA, 1981). To understand the motivation and the result soon after the initiation of the CPAC's activities, these three publications are necessary reading for the analysis and evaluation of the management of knowledge generation, either the suggested or unspoken interest of each researcher, such as the explicit knowledge regarding the interest of organizations and 'caretakers.'

The performance of Brazilian agricultural production, with the exception of coffee and sugarcane, which are commodities that did not entirely come under the jurisdiction of the Ministry of Agricultre, depended on what little nutrients the soil had to offer, as well as rain and heat. At the time, coffee and sugarcane producers, and much later, cacao producers, did not consider themselves as farmers. A farmer was considered a second-class, if not a third-class citizen. The farmer was akin to *Jeca Tatu*,[5] a slave and a colonist. Cattle husbandry, also extractive, was undertaken by big families owning large fields and woodlands.

This paradigm was solved upon researching and subsequently proving that the Cerrado soils that were considered unfertile could, when properly managed, be transformed into soils with reasonable quality suitable for any agricultural production in a rural environment. Former slaves became coffee producers and settlers, while other entrepreneurs opened sugarcane mills; agriculture and cattle husbandry needed to evolve and adapt to the new forms of labor. It was necessary to absorb and adopt new and unpracticed technologies.

Brazil was going through difficult times in the 1950s and 1960s. In 1950, Brazil processed less than 1 percent of its agricultural products; between 1965 and 1974, the production of five key crops (rice, beans, potatoes, cassava, and wheat) fell by 18 percent; in this same period, the prices of these five products increased by 108 percent. In 1973, the worldwide oil crisis exacerbated this scenario, along with the scarcity of currency exchange and the need to import food; the second oil crisis in 1979 confirmed the need to expand the country's agricultural frontier by using Cerrado soils for the production of bio-energy, along with food, beverages, fibers, essential oils, and other products.

The first CPAC Research Program (1976/1977) addressed these situations, pursuing solutions aimed to minimize or eliminate hindrances and problems created by the rainy, dry and/or *veranico*[6] seasons. These solutions came through the introduction of perennial crops, forages and pastures, as well as irrigated crops, not only as alternatives to upland rice cultivation and pastures, which were dominant at the time, but also for the rational and organized use of resources throughout the whole year.

As for soil-climate-plant relations, the CPAC tried to introduce the results of research of the Procal Program, mainly from *Cerrados* of São Paulo, and of the Tatu Operation Program of the northern Rio Grande do Sul state. As in these programs and the respective soils, in the soils of the CPAC's core area the interaction between liming and phosphate fertilization gave positive results.

The entire agricultural performance of the Cerrado region occurred to a great extent because of knowledge generation and technologies developed by the CPAC in interaction with the EMBRAPA system, with local organizations and institutions from other countries, as shown in the summary below. The development also relied on farmers and their families, who grew alongside the research.

## (2) Technical-scientific production generated and/or validated by the direct action of the CPAC[7]

### *(a) 1975–1984*

The first ten years of the CPAC's life were very promising in terms of the development of its technical-scientific team, the programming of experimentation and research, infrastructure bases, and laboratory support. During this ten-year period, four major events took place which introduced research and development into the Cerrado region: (1) IV Symposium on the Cerrado, in June 1976, with focus on the bases for agricultural utilization; (2) International Seminar on Tropical Soils, in February 1978; (3) V Symposium on the Cerrado, in February 1979, with discussions on the use and management of soils; and (4) VI Symposium on the Cerrado, in October 1982, with focus on savannahs: food and energy. This last event initiated the pursuit for international cooperation. It was in this period that the JICA–CPAC cooperation agreement

*Figure 7.3* Photos of Cerrado types ('clean' field, 'dirty' field, Cerrado and Cerradão) and crop/cattle husbandry/ forest

*Source:* Authors.

began (1977–1994), among other important agreements executed with CIAT, IICA/Bird, Tropsoils, Orston, and Cirad.

## (b) 1985–1994

The results of the work conducted in this period are reported in three technical reports of 1985–1987, 1987–1990, and 1991–1995, in addition to publications in other media such as specialized magazines and JICA–CPAC cooperation reports. The most important event of this period was the VII Symposium on the Cerrado, with the objective of analyzing Cerrado utilization strategies, in April 1989.

From a model focused on a few experimentation and research projects, EMBRAPA decided, from 1981/1982 on, to adopt a diffuse model of agricultural research so that each assay, test or experiment would constitute a project assigned to a researcher. This measure made not only the adoption of a systemic focus of the corporation difficult, but also the coordination of the CPAC, both internally and externally, with its collaborators. The establishment of the National Research Program (PNP) reduced the difficulties, and the Center started to operate the three above-mentioned PNPs.

## (c) 1995–2010

In the third period, three events are noteworthy: (i) VIII Symposium on the Cerrado, with a focus on biodiversity and on the sustainable production of food and fibers, in March 1996; (ii) Symposium on Ecology and Biodiversity, in June 2002; and (iii) IX National Symposium on the Cerrado, with the theme: challenges and strategies for the balance between society, agri-business and natural resources, in 2008.

The Center made three principal technological contributions to the development of Cerrado agriculture: (1) rhizobium bacterium, for nitrogen fixation in seeds, when applied in soybeans – positive results are shown, as well as in beans, peanuts, peas and lentils. This technology reduced the costs of nitrogen application, saving approximately US$5 billion/year, which made Brazilian soybeans the cheapest in the world; (2) direct cultivation, generating savings on fertilizers and machinery equaling approximately R$1 billion/year; and (3) genetic improvement to adapt soybean varieties, starting in the late 1970s, when the plants grew less than 20cm tall and productivity was around 1,000 kg/ha. As a result, average productivity increased to 3,000 kg/ha.

This set of three technologies and other practices made it possible to obtain the current level of soybean productivity for commercialization in the Cerrado. As of today, the production of 4,920 kg/ha, and the national average of 3,012 kg/ha, are greater than the productivity of either the United States or China. Moreover, the success of soybeans as the leading crop in development influenced other segments.

## 7.3 Japanese cooperation

In September 1974, the Japanese Prime Minister, Kakuei Tanaka, visited Brazil and signed a joint statement with President Ernesto Geisel. This motivated the dispatch of a Japanese mission to Brazil in February of 1975, and another in September of the same year. In March 1976, a third mission visited Brazil with the purpose of formulating the bilateral cooperation register. The first cooperation agreement was signed on September 30, 1977.

At the time, the Minister of Agriculture, Forestry, and Fisheries of Japan created a department in the Tropical Agriculture Research Center dedicated to the interests of Japanese researchers and technicians, including the Brazilian Cerrado. In July 1977, this Center published 9 articles out of the 26 that were part of the III Symposium on the Cerrado.[8]

The official visit of President Ernesto Geisel to Japan in 1975 corroborated the growing interest of both countries, consolidating what had been published in 1972 as the Annals of the Second Brazilian Meeting of the Cerrado.

The Japan–Brazil cooperation agreement had the strategic vision that agricultural research would be necessary within the scope of Cerrado development. The CPAC was chosen to carry out this research, entering into a technical cooperation agreement with the Japan International Cooperation Agency (JICA), with the commitment that state research institutions, such as EPAMIG (MG), EMGOPA (GO), and EMPA (MT), would be included and would benefit, in addition to future equipment donations by the Japanese government.

An unbiased and internal evaluation of the CPAC, regarding the process of Japanese–Brazilian technical cooperation, has always highlighted cultural and language differences between the two countries, regardless of the fact that Brazil has a significant number of people of Japanese descent – possibly the largest in the world. Upon the first team's arrival, the picture was completely different. In terms of personal relationships, a close bond was observed, which substantially facilitated the technical partnership.

The Brazilian side understood that it was natural for the newcomers to be surprised by the dimensions of the country and of the Cerrado.

Conversely, CPAC research projects were just beginning, and many of the observations ultimately became the object of 'reconfirmation' by Japanese researchers. In other words, some tests that gave rise to doubt were repeated.

This fact shows how well the two partners bonded and shared technical-scientific opinions about problems and possible solutions for the development of the Cerrado. When contradictions arose, it was always positive and healthy from a scientific and technological standpoint.

The four cooperation project reports are testimony to this connection, since Brazilian and Japanese teams, as well as other institutions and countries, took part in the same research programming of the EMBRAPA system. This was one of the most important reasons that important results were obtained in such a short period of time.

The general evaluations of Japanese cooperation and of cooperation with other countries and institutions, especially international and inter-governmental institutions, are not necessarily comparable. In the same period, the CPAC had the cooperation of the International Center of Tropical Agriculture (CIAT), in pastures and forages, and of the Tropical Soils Program (TROPSOILS) with the Universities of North Carolina/ Raleigh, Cornell/Ithaca, Hawaii, and Texas (A&M). In this coopera-tion agreement, the General Chief of the CPAC, Elmar Wagner, was a member of the Board of Trustees of CIAT/Cali-Colombia for six years and, during the same period, a member of the Board of Directors of TROPSOILS. Orston and Cirad, both from France, also collaborated on specific research topics.

The cooperation with JICA was unique, an example of the objec-tive of collaborating for the institutional building of the CPAC and of some of its most direct relationships; in addition, this was its reason for being: a higher development proposal involving the interests of both countries. With that, this cooperation with JICA is defined and has a basis similar to the cooperation of the International Bank for Reconstruction and Development (IBRD), of the World Bank (WB), the technical component of which was conducted through the IICA. However, that was regarding a bank and financing, not about the donation of knowledge, techniques/methods and equipment, such as the case of Japan. In addition to donations, the cooperation of technical teams from Japanese universities and the Japanese Ministry of Agriculture, Forestry, and Fisheries was essential, as well as visits by CPAC staff and other research units to Japan, for training and qualification, and also to understand the culture and socio-economic background of the cooperation partners.

The cooperation in agricultural research between Japan and Brazil through JICA and the CPAC was enshrined in cooperation agreements. The first agreement was signed on September 30, 1977, and covered three teams of Japanese researchers who worked at the CPAC and in other states, from 1978 to 1980, from 1980 to 1983, and from 1983 to 1985, respectively. The second agreement was signed on August 30, 1987, with the participation of three more teams: from 1987 to 1990, from 1990 to 1992, and in 1994/1995.

The donation of laboratory equipment by Japan had an impact on the quality not only of analyzes and image processing, but also of learning and the theses or dissertations produced in support of academic degrees. The total amount donated in equipment and technical-scientific instruments was estimated at US$8 million.

*Table 7.2*  General distribution of Japan–Brazil/JICA–CPAC cooperation (two agreements and five teams of Japanese researchers (E1 to E5) with corresponding number of researchers from the CPAC)

| CPAC chiefs | Period | JICA leaders | Period | Number of researchers from JICA | Short-term Japanese consultants | Number of researchers from CPAC |
|---|---|---|---|---|---|---|
| Elmar Wagner | 1976–1984 | E1 – Yoshiro Sakurai | 1978–1980 | 8 (1 admin.) | 5 | 14 |
|  |  | E2 – Tamotsu Ogata | 1980–1983 | 7 (1 admin.) | – | 14 |
|  |  | E3 – Tamotsu Ogata | 1983–1984 | 7 (1 admin.) | – |  |
| Raimundo Pontes | 1984–1985 | E3 – Tamotsu Ogata | 1985–1985 | 7 (1 admin.) | – | 18 |
| Guido Ranzani | 1985–1986 |  |  |  |  |  |
| Wenceslau Goedert | 1986–1988 | E4 – Bunkishiro Watanabe | 1987–1988 1988–1989 | 13 (1 admin.) | – | 24 |
| Carlos M.C. Rocha | 1988–1989 |  | 1989–1990 |  |  |  |
| José R.R. Peres | 1989–1993 | E5 – Tadashi Morinaka | 1990–1992 | 12 (1 admin.) | – | 24 |
| Jamil Macedo | 1993–1996 | E5 – Toru Kubota | 1994–1995 | 4 | – | 4 |
|  |  | Publication of the Technical report of teams E4 and E5, 1987–1992 (Brasília/DF, 1994). |  |  |  |  |

*Source:* JICA–CPAC project reports.

## Results of JICA–CPAC cooperation

The results of the JICA–CPAC cooperation are principally included in four project reports, mentioned in the bibliographic review as cooperation agreements (partial reports, technical-scientific works, *Considerations on the agricultural development in the Cerrado* and the Project's technical report), published between 1980 and 1994.[9]

In the first period, considerable efforts were made by specialists in the selection and assembly of equipment and instruments, mainly laboratory equipment. There was important scientific and technological contributions in the fields of plant pathology (resistance of *Stylosanthes spp.* to antracnosis, a mosaic virus in cassava, watermelon, and cucumber) and entomology (biology and control of *Elasmopalpus lignosellus*, bedbugs, in the cultivation of soybeans). Research into soils (radicular system, soybean development, analytical methods for Al and Mo in parts of the plants, effects of the application of lime and phosphates on soybeans), phytotechnics (cultural systems of soybeans, wheat, the control of weeds, soil compaction, plowing, and root development), and agrometeorology (agrometeorological research and metric techniques) complemented this initial study phase.

The second team of specialists continued the activities started by the previous team, and expanded the range of actions and activities within the field of phytopathology/entomology (to annual crops and pastures) and to new fields, such as management and mechanization, soil and water, and natural resources (techniques of remote sensing in vegetation burning and land use, evaluation of Landsat images, spectral signatures in varieties of soybean, spectral responses in forestry areas).

The third team not only worked on scientific issues, but also promoted a general evaluation of the stage of Cerrado development (more than 6 million ha utilized) and possible future needs due to the expansion of the agricultural frontier in the Cerrado towards the states of BA and MT, with their different natural and socio-economic conditions. The title of its report, *Considerations about the agricultural development of the Cerrado*, suggests this mix of research actions, results, and potential problems.

The evaluation of the results obtained by the first three teams allowed the Mission Leader, Tamotsu Ogata, to recommend that the agreement be renewed and expanded to the states of BA and MT.

The *Technical report of the Japanese–Brazilian project of cooperation in agricultural research in the Cerrado* recorded the actions and results of the two teams covered by the second cooperation agreement (see Table 7.3). The report mentions 136 million ha of land farmable for the production

*Table 7.3*    Technical-scientific production of JICA–CPAC cooperation

| | 1980–1989 | | 1990–1999 | | 2000–2004 | |
|---|---|---|---|---|---|---|
| **Publications** | Total | 1981 | Total | 1994 | Total | 2000 |
| | 39 | 19 | 60 | 36 | 24 | 21 |

*Source:* JICA–CPAC project reports.

of food, fibers, and energy. At the time, in 1994, with only 23 percent of its potential area utilized, the Cerrado accounted for approximately 28 percent of national grain production and 20 percent of national coffee production, and had 40 percent of the national cattle herd.

The activities of research and experimentation consisted of work in plant science (analytical studies on soybeans, tolerance to aluminum in wheat and soybeans); in plant pathology: entomology (insects that damage soybeans, natural enemies in the control of insects, ecology of bedbugs of rice stems, occurrence and biology of *Diatrea saccharalis* in rice, studies on *Entomogenous fungi* in the microbiological control of pests), and plant pathology (classification of races of soybean mosaic virus, and isolation of leaf bacteria, rice diseases, white mold of irrigated bean plants, mosaic virus in bean plants); in agricultural mechaniza-tion (laboratory automation systems and fuel consumption of different equipment); in natural and socio-economic resources (analysis of mete-orological and geographic variations, behavior of crops in sustainable growth, and economic conditions, notes about econometric methods); and in soils and water (supply of nitrogen, nitrification, time to irri-gate beans and peas, drainage rates, plowing and compacting methods, plant root problems, organic matter, new inoculation techniques, chromatography).

Within the scope of the technical cooperation, the following activities were developed: (i) a system for the acquisition of agricultural machinery data; (ii) biological control of the rubber tree lace bug; (iii) a catalog of images relating to agriculture in the Cerrado; (iv) the Japanese transla-tion of the book *Soils of the Cerrado: Technology and management strat-egies*; (v) Joint Seminar of Covenant Institutions, in Brasília (October, 1989), and (vi) Seminar on the Progress of Agronomic Research in the Cerrado Region, in Cuiabá/MT (October, 1991).

Kubota (1996) reports that up until that time the project had already involved 77 Japanese researchers and technicians, and 62 Brazilian coun-terparts, and spent an amount totaling approximately US$5.2 million in equipment and facilities.

The total number of technical-scientific publications directly resulting from the cooperation agreement, 1980 to 2004, was 123, of which, 39 were published in the period between 1980 and 1987, with most published in 1981, 60 publications between 1991 and 1999, with a larger number published in 1994, and 24 publications between 2000 and 2004, with a larger number published in 2000. [10]

Japanese–Brazilian research cooperation covered the first 20 years of the CPAC's existence (1977–1995) and was one of the most positive aspects of the cooperation between both countries.

## 7.4    Activities of diffusion and rural extension

The Project of Agricultural Research and Bilateral Cooperation in Agricultural Research aimed to incorporate most of the Brazilian territory in the production of goods and services for the development of the country and cannot lack a powerful technology transfer scheme and adoption for stakeholders such as businessmen, rural producers, and for professionals in technical assistance and rural extension. Society, in general, needs to follow the progress of such projects.

EMBRAPA, since its establishment, had developed the concept of technology diffusion based on the idea of diffusing the knowledge created and/or validated directly to the Rural Extension Services, to technical assistance professionals, and to the rural producers themselves, as well as to society. The means and procedures to work accordingly are surveyed and explained at the same time when these mechanisms are evaluated for their performance in terms of functions and results produced.

The concept of diffusion comprises principles and definitions related to the following: (i) information – explanation, provision of data, notes, arguments; (ii) communication – participation, information, notice, transmission, connection; (iii) transfer – act or effect of transferring, removal, assignment; (iv) training – act or effect of training, exercise; (v) qualification – to train and certify, persuade, habilitate; and (vi) IT – information technology – in cybernetics, qualitative factor that designates the position of a system, possibly transferable to another system. Each of these elements is developed by its own procedures or processes.

CAMPO was an institution that had undergone these processes, being the subject and object, in its function as a source of research, knowledge and technology, and as a user of research results. Work articulated with the CAMPO is aimed at resulting in a precise evaluation of the results obtained through these mechanisms. The transfer and the adoption of technologies by producers associated with the CAMPO was to be evaluated through the performance of different settlement projects over time, in terms of productivity and of production quality, and the CAMPO itself was to evaluate what it had obtained from the cooperation, not only from CPAC, but also from other institutions.

During the period of JICA cooperation, the CPAC kept a Rural Extension Office, connected to the DF Rural Extension Service, which fulfilled the function of informing, communicating, transferring, training, and qualifying the members of production chains, including rural extension professionals – mainly producers within the Federal District and of the Geo-economics Region of Brasília.

In the first 20 years of the CPAC's existence, the period covered by the Brazil–Japan cooperation, the concept of technology diffusion underwent changes in principles and definition, until the moment of the creation (1981/1982) of the Technical-Scientific Communication Coordination Office, integrating the Sectors of Publication, Diffusion, Information and Documentation (Library), Journalism, Training, and Distribution and Sales. Thus, for the first time, a whole articulated set of related activities complementing the research was organized.

At the CPAC, technology diffusion was understood as a two-way transfer: (i) dissemination of knowledge; and (ii) indispensable feedback for constant adjustments and readjustments to the direction of the research programs. The diffusion of technologies had been exercised since the first year of the CPAC in the form of 'discussions' about the opening of the Cerrado's natural vegetation, field trips, technical visits, seminars, field demonstrations, and publications, culminating in the IV Symposium on the Cerrado.

The utilization of 'technological packages' represented a great contribution to the dissemination of knowledge among rural producers, extension professionals, and researchers. It worked as a 'funnel', where the input for information from producers was represented by a vector bigger than the others, extension professionals and researchers. At the output end of the funnel, production systems were put together: S1 low-grade technology; S2 medium-grade technology; and S3 high-grade technology, considering the knowledge available at the time. As time went on, the research inverted the vectors: researchers got to know more and became the larger contributing vector, while the producers' vector became smaller. These production systems were published as small booklets or textbooks.

The year 1977 was considered the Year of Diffusion, with 13 meetings about technological packages, with the participation of 210 producers and extension professionals, and of 197 researchers. In the following year, incentives were given to produce technical communications and to support producers' demands for technologies, while maintaining other activities. Most of these activities included assistance to visitors, producers, representatives, schools, foreign scientists, and missions, such as from the Superior School of War, with 126 participants. The visit of the British Prince Charles in March 1978 was a highlight. In February 1978, the Technical Meeting on Tropical Soils was held with the participation of the Universities of North Carolina, Cornell, Puerto Rico, Ghana, and Peru.

In 1979, attention was given to the relations with the Brazilian ATER System (SIBRATER), with training and expansion to MS, MT, and MA (Balsas). In February 1979, the V Symposium on the Cerrado was held.

In the following years, actions focused on the diffusion of technologies, training, and continuous qualification in technical-scientific divulgation, with unscheduled visits from over 200 persons.

In these 20 years, the Center did not continue the same activities in terms of the diffusion of technology, using all denominations possible, but it did continue to give priority to the diffusion of technology and knowledge, as well as to the internalization of externally generated knowledge.

The CPAC's extensive experience in this field makes it a reference for technical-scientific experimentation and research, as well as for the analysis of institutions from Brazil and abroad.

## 7.5   The CPAC from global perspective

Since 1980, the CPAC has been working at the international level, examining tropical soils around the world and topics related to savannahs. Two international symposiums have been held, whose outcomes have been published as the Proceedings of the events (1st International Symposium on Tropical Savannahs, March 24–29, 1996, Brasília/DF; and 2nd International Symposium on Tropical Savannahs, 2008, Planaltina/ DF). During the second symposium, associated with the IX National Symposium, a workshop entitled 'Savannah: demands for research' was held, in which 12 working groups discussed the following issues: (1) savannahs in the world; (2) characterization, conservation, and use of biodiversity; (3) characterization, conservation, and use of soil and water; (4) agricultural and forestry production; (5) impacts of production systems and mitigation strategies; (6) agricultural commodities and socio-environmental appraisal; (7) biotechnology, transgenic and biosecurity; (8) agro-energy; (9) alternative and diversified systems of production; (10) family agriculture; (11) precision agriculture, agro-environmental zoning and modeling; and (12) public policies and worldwide perspectives for savannahs.

As a result of this workshop, savannahs came to be "characterized by the vegetation found in tropical and sub-tropical regions, which is under a long dry period, receiving more rainfall than desert regions, and physiognomically having sparse trees and shrubs, or sparse groups of trees and shrubs, with a very pronounced graminoid stratum" (Allaby, 1988; Art, 1998).

According to the *Encyclopedia of the Biosphere* (2000), savannahs cover nearly one-fourth of the earth's land area. They are found on all continents, with a major presence in more than 30 countries. They have a long history of human use, and currently house approximately one-fifth of the world's population. Recognized by the general public as a natural environment where wild animals live, savannahs receive little attention from research as an environment with sustainable development potential for mankind.

The main determining factor of this ecosystem is the annual rainfall distribution pattern, with two clear seasons: dry and rainy. The amount of rainfall and the duration of these seasons condition the type and volume of vegetal coverage, the type of predominant fauna, and consequently the degree of human use and occupation.

Most savannahs are located between the tropics of Cancer and Capricorn, and these are called tropical savannahs. They have huge potential for agricultural production (food, fibers, energy, etc.), in addition to other activities of social, economic, and environmental interest.

Among the main challenges mankind will have to face in coming generations include issues closely related to agriculture, particularly food production; it can therefore be predicted that there will be greater use of the tropical savannahs. The others challenges include water supply, energy production, environment conservation and poverty reduction. The solutions to these problems greatly depend on the sustainable utilization of natural resources and agricultural production activities.

The amount of fresh water on the planet is finite, and its scarcity is causing increasing concern. It is already a limiting factor for the development of many countries, especially those in arid and semi-arid areas, in which part the savannah ecosystem is included. In addition, it is important to remember that the sources of the main water courses of the planet are located in regions covered by tropical savannahs, and the intensive use of them should not jeopardize the flow and the quality of this water.

The demand for food, in quantity and in quality, will increase proportionally with the increase of the population. Meeting this demand poses a complex equation, including an increase in productivity in areas currently explored and the expansion of new areas, as is already taking place in the savannahs of some countries, such as in Brazil, Zimbabwe, and Thailand.

The trend of the forthcoming energy matrix is to increase the participation of sources from agricultural-based sources (biomass, alcohol, biodiesel, etc.), which represent more pressure on the use of tropical savannahs. In summary, the solution to the problems concerning water, food, and energy will involve an intensification of the use of tropical savannahs, increasing pressure on the environment, and as a result, increasing the risk of degradation and desertification, especially with regard to more fragile environments.

The eradication of poverty in the world is certainly the most comprehensive and complex challenge because it has its origin in cultural and political aspects, present in various countries where tropical savannahs perform an important role, especially in the African continent. Briefly speaking, the close relation between the sustainable development of society and the use of tropical savannahs is evident. Furthermore, it is possible to predict that this system will perform the most important role, when compared to others,, as discussed further below.

The geographical distribution of tropical savannahs shows that this ecosystem occupies most tropical and sub-tropical areas, representing approximately 50 percent of the African continent, and a significant percentage of South America, Asia, and Oceania. As Table 7.4 shows, this ecosystem occupies a significant area in more than 30 countries, approximately 2 billion hectares, while the Brazilian Cerrado represents only approximately 10 percent of tropical savannahs.

Tropical savannahs are stable ecosystems where the main ecological determinants are water and nutrient availability. Although highly diverse, the natural resources of this ecosystem can support more intense use than those being currently exploited, thus having great agricultural potential.

## The CPAC's tecnhical cooperation in Mozambique

The essential purpose of the CPAC's technical cooperation in Mozambique is to contribute to the regional development of the Nacala Corridor in Mozambique. The objective is to develop knowledge and generate agricultural technologies adapted to the region by strengthening the

*Table 7.4* Distribution, location, denomination, dimensions, and population of tropical savannahs

| Continents/ formations | Main countries | Local denomination | Area (km² 10⁶) | Population* (%) |
|---|---|---|---|---|
| South America | Brazil | Cerrado | 2.4 | 6 |
| | Colombia and Venezuela | Lhanos | 0.6 | |
| Africa (West) | Senegal, Guinea, Ivory Coast, Mali, Ghana, Benin, Nigeria, Cameroon, Chad, and Central Africa | Savannah | 5.0 | |
| Africa (Central/ South) | Angola, D.R. Congo, Zambia, Malawi, Zimbabwe, Mozambique, Namibia, Botswana, and South Africa | Miombo | 4.5 | 13 |
| Africa (East) | Ethiopia, Somalia, Uganda, Kenya, Tanzania | Savannah | 2.5 | |
| Asia | India, Burma, Laos, Thailand, Vietnam, and Cambodia | Savannah (Dipterocarp) | 2.5 | 60 |
| Oceania | Australia | Savannah | 2.0 | 1 |

*Note:* *Proportional population of each continent in relation to the world population.
*Source:* Authors.

capacity of agricultural research and diffusion at the Institute of Agrarian Investigation of Mozambique (IIAM), in its experimentation stations of the Northeast Zonal and Northwest Zonal Centers. The objectives of the technical cooperation project are:

(a) To strengthen the operational capacity of the Northeast and Northwest Zonal Centers of IIAM;
(b) To evaluate the socio-economic conditions and develop methods and criteria for the appraisal of the socio-environmental impacts of the use of new technologies in the Nacala Corridor;

(c) To identify and evaluate the natural resource conditions for the practice of agriculture in the Nacala Corridor, and to generate and provide technologies for sustainable regional development;

(d) To generate and provide knowledge and technologies appropriate to the growth of agricultural crops and animal production in the Nacala Corridor; and

(e) To develop, validate, and implement agricultural technologies in selected Demonstration Units.

The project proposal has five components:

(1) Strengthening of the physical and institutional infrastructure of the Northeast and Northwest Zonal Centers of IIAM;

(2) Evaluation of socio-economic conditions and development of methods for the appraisal of socio-environmental impacts;

(3) Evaluation of natural resource conditions in the Nacala Corridor region. Development and provision of technologies for sustainable agricultural use;

(4) Support for the research and development of technologies appropriate for agricultural crops and animal husbandry in the Nacala Corridor; and

(5) Installation and implementation of Demonstration Units in the vicinity of the Lichinga and Nampula Experimental Stations.

The implementation of these components is under the responsibility of some of EMBRAPA's centers related to those specific issues.

Although there is a difference in territory dimension, and certainly other social, cultural, political, and economic differences, as well as of natural, technological, and production resources, there is the strong possibility of transferring research management technologies, information, results, and knowledge to the benefit of individuals, organizations, and stakeholders in the agricultural production and agri-business sectors.

Mozambique has private sector institutions, providers of inputs and services, institutions in the public sector, various services provided to society, international and intergovernmental public sector institutions, and third sector institutions. There are several institutions cooperating with the country, and there should be various institutions included in the third sector.

The focus on environmental issues concerning natural resources shall result in the adoption of a work scheme which has a framework, on one side, of the public/collective interest assigned to national and international sectors, and on the other side, the private interest of individuals, families and organizations, as well as of the third sector. At the top of this framework, there are the production management models, and at its base, environmental management systems. Inside the framework, the organization of the three segments of interest, research and development attribution is as follows: (i) social resources; (ii) natural resources; and (iii) technological and production resources. These three components may be represented by overlapping circles defining interacting areas.

Among the above-mentioned three components, technology is the most important because of its fundamental role in guiding social intervention in the environment in order to help define development policies and the construction of production management models. Policies and models, among other results, comprise the respect given to social diversity, the maximization of manpower, and the rational use of natural resources.

The harmonic interaction between the three components should result in the construction of a sustainable agri-business for each country or region, as is already taking place in some countries, especially in those with a stable political and institutional situation and those investing in research.

## Appendix: Technical-Scientific Production Generated and/or Validated by Direct Action of the CPAC

### 1   1975–1984

*(a) PNP resources evaluation*

In this first period, the project was dedicated to the pursuit, collection, and methodological treatment of information on existing knowledge of the Cerrado's natural and socio-economic resources. The first major challenge was to work out the dimensions of the Cerrado's biome, which was defined by vegetation physiognomy parameters as 120 million ha, while according to other criteria, as 240 million ha. For a long time, the average of the two values, 180 million ha was used, of which at least 50 million ha were farmable.The first year of activities was focused on a survey and classification of the soils in the Federal District soils, interpretation of remote sensors in the DF area, and a survey of entomological fauna in the Cerrado area, in relation to the main crops (rice, corn, wheat, soybeans, and cassava). In the second year, an agreement with the Brazilian Institute of Geography and Statistics (IBGE) was made for the regionalization and agro-climatic zoning of the Cerrado. In addition to this, studies on water resources, which started being promoted in May 1977, produced a 'Work Table about the Water Problem in the Cerrado'.

In 1978, the EMBRAPA/CNPq agreement was signed, allowing activities in the University of Brasília's Flora Program, and EMBRAPA/INPE, for environmental appraisal. In the same year, a collaboration program for the evaluation of Cerrado areas for agro-forestry-pasture purposes was started with CIAT. Also in that year, research into forests, which at the time was the responsibility of the Ministry of Agriculture, became the responsibility of the CPAC. At the time, the project produced the first water and climate balance of the Cerrado, and the first preliminary list of native species bearing economic potential. In 1980, the Project of Inventory and Evaluation of Natural and Socio-economic Resources (PNP Evaluation) was already operational, with ten research projects, and was organizing the Environmental Databank in 143 weather stations, in addition to climatic, phytogeographic, and pedologic maps.

In the early 1980s, through various approximations, the PNP Evaluation concluded that the Cerrado's area was 2,037,600 km², or 203.7 million ha.

Volume 8 of the CPAC's report for 1982–1985, as the result of the change in the scheduling scheme, summarizes the work developed by the PNP Evaluation in 21 study and research projects.

Among the countless contributions and results of the PNP Evaluation, the following are noteworthy: (i) Environmental Databank – with analyzes and the expansion of information, especially that of rainfall data and patterns; (ii) ecology of the Cerrado, representativeness of CPAC's area – in functions of Cerrado's different environments, geographic limits, typology of native species, types of vegetation/phytophysiognomic, plant phytosociology, soil/plant relations – to the utilization of native and forage species; (iii) phenology and physiology of the germination of species of interest; (iv) evaluation of current production systems, including socio-economic evaluation; and (v) remote sensing techniques, radiometric analyzes, photographic processing of pastures, digital processing of images, and radiometry of targets.

These studies served three major objectives: (i) to objectively guide the scheduling of CPAC's and its partners' research; (ii) to instruct and to guide the ATER system with respect to the potential and available technologies for transfer; and (iii) to guide interested institutions and 'caretakers' to settle in the Cerrado region.

*(b) PNP utilization*

Upon working out the soil-climate-plant relations, scientists soon realized the need to work on basic research issues to expand the spectrum to soil-water-plant/animal-atmosphere. With this procedure, the project team began to interact with other scientific centers – with universities at home and abroad, through cooperation and exchanges, and on Master's and/or Doctor's degrees thesis programs.

The organization of research trials, tests, and other experiments took place mainly as a process of scrutiny, while focusing on the fundamental problems detected in various sources: (i) soil acidity; (ii) nutrient deficiency; (iii) water shortage; and (iv) phosphorus fixation. However, the initial focus was on issues related to the soils of the Cerrado, such as: corrections of acidity and toxicity, microbiology, organic matter, management of fertilizers, chemical and mechanical soil amendment practices, influence of soil humidity, of rainfall/simulations, and of erosive processes.

The identification of low soil fertility stems from the high power of phosphorus fixation, caused by high saturation of aluminum, low cation exchange capacity, and widespread deficiency of nitrogen, potassium, magnesium and zinc; the correction of soil acidity and the management of nutrients allowed the identification of various alternatives for the same problem. These alternatives were supported in terms both by phosphorus adsorption by soil and absorption of phosphors and other nutrients by plants.

As in other regions in Brazil with acidic soils, in the Cerrado the interactive effect of lime silicates and phosphates applications, ways of gradually correcting soils – doses and depths of incorporation – and their residual effects were tested. Recommendations regarding these results were soon sent to technical assistance and extension services, to producers, and to financing agents, particularly to the Central Bank, to be included as investment items in loans, along with dolomitic limes, sources of phosphorus, especially the soluble ones for annual crops, and the rock phosphate for pastures and other perennial crops.

In the extensive task of screening species and varieties with regard to the high-levels of exchangeable aluminum and the correction treatments, positive results were obtained, resulting in increased penetration of root systems, and thus increased absorption of nutrients and water in the subsoil. The most evident result for the neutralization of exchangeable aluminum was the application of limestone and, later, of gypsum. Different sources of phosphorus in different soil conditions, of differing plants/varieties were tested, and reacted in different ways.

When there is the intention to use native Cerrado soils, it is advantageous and absolutely necessary to recuperate those soils in order to increase the level of available phosphorus, until the next critical level, which can be done immediately or gradually.

The Cerrado's soils are generally very poor, lacking most nutrients, and have only medium levels of organic matter. In experiments with different sources of phosphorus, upon studying mechanical, chemical and biological means, the efficiency in the mycorrhizic association and applications of phosphorus pentoxide, $P_2O_5$, only in soils infected with spores was observed. In 1980, after four years of research, the CPAC already had recommendations for the acceptable management of Cerrado soils. In order to increase soil fertility (chemical recuperation) and manage Cerrado soils, methods of analysis, levels of phosphate fertilization, evaluation of sources, effects of phosphorus and lime interaction, effects of acidity in the efficiency of natural phosphate, mycorrhizal and efficiency in the absorption of phosphorus, organic matter and micronutrients, were all elaborated. There was also significant progress on the topics of biological nitrogen fixation, water deficiency due to mechanical practices, irrigation and drainage, and management and conservation of soils, helped by the installation of a rainfall simulator.

The handling of water deficiency and the effects of water stress resulted in the need to evaluate the depth of lime incorporation and application of water during the rainy season's dry spells – *veranicos* – through different irrigation methods.

The likelihood of erosion becoming a problem, due to more intensive use of the soil, was anticipated and several experiments were established for this purpose and further understanding.

Between 1976 and 1977, work began on the occurrence and behavior of microorganisms, mainly on those responsible for atmospheric biological nitrogen fixation and cellulosic decomposition. Soybeans were the main crop of interest for these studies, and showed the inexistence of rhizobium strains (*Rhizobium japonicum*) in the soils studied. With that, assays of several *Rhizobium spp.* strains, with different levels of inoculation, including pelletized and non-pelletized commercial inoculants, were initiated and their results evaluated.

In 1977/1978, a series of studies on the occurrence of insects, pathogens, and invaders was published. These included spittlebugs (*Homoptera: Cercopidae*) in *Brachiaria decumbens* pastures, wheat and citrus lice (*Leptopharsa heveae*) in cassava, and the rice borer insect. The research on this issue also resulted in the surveying of various species, which can be found in the Entomologic Museum of the CPAC, housing approximately 7,500 specimens.

In the same year, a survey of phytoparasite nematodes associated to 44 species of cultivated plants was started. The research concluded that control can be obtained through soil fumigation using biological methods and farming practices. The only bacterium studied in this period was *Xanthomonas manihotis* in cassava.

In the third year, invasive plants (weeds) occurring in the south of the country were found, in addition to the regrowth of native plants behaving like invasive species. Studies of chemical, mechanical, and biological control methods in the following year (1978/1979) suggested that chemical and mechanical control had the same efficiency, the latter being superior if considering economic and environmental aspects.

In 1980, a research program on agricultural mechanization began, with a focus on the high consumption of fuel and premature weathering of machines and attachments. In the first phase, the development of machines with solutions for specific problems was pursued, with the elaboration of a prototype grass seeding machine, an innovation patented by EMBRAPA; a bendable tilting cart; and a garlic planter, both filed for patent. Between 1982 and 1985, a furrower coupled to a seeder-fertilizer machine was developed, for the implementation of furrow irrigation crops.

In 1981, studies on the management of wet areas began (water movement in porous mediums, spacing between drains, and management of annual crops). From 1982 to 1985, the predominant soil types in 4,562,445 ha of the 11,920,000 ha estimated as existing in the water basins of the São Francisco, Paranaíba, Araguaia, Tocantins, and Alto Paraguai (Pantanal) Rivers, were characterized.

Later in the 1980s, research began on the supply of energy to agricultural properties from renewable sources, using bio digesters, solar panels, and windmills.

*(c) PNP production system*

By the end of the first year of activities, the experience acquired by the CPAC's team and the results obtained from research were substantial, along with the contacts within the agricultural community, for the agricultural development of the Cerrado region, where the following issues had been identified: (1) it was necessary to stabilize the yield of the traditional (rice/pasture sequence) production system, which was in decline; (2) due to the low natural fertility of the soils, it was necessary to promote their initial improvement (recuperation) through liming and corrective fertilization, mainly using phosphates, considering that the parceling of this practice would allow the farmer to obtain much more experience with crops; (3) the improvement of the soils required the introduction of crops with a higher return potential, such as soybeans, corn, and wheat; (4) in animal production – beef cattle – food scarcity in the dry season (May to October) necessitated supplementing the feed with hay; and (5) owing to the long dry season, with reduced agricultural activity causing discontinuity, it was necessary to create alternatives to irrigated and perennial crops (coffee, citrus, forestation).

In 1976, the IV Symposium on the Cerrado, bases for agricultural utilization, gathered and consolidated these observations. Also, it projected new paths for the agricultural research of the Cerrado upon impending issues such as ecology, zoning of climatic aptitude for commercial crops, mineral resources, flora, socio-economic resources, microbiological balance of soils, suggestions for research on water deficiency, a model for supplementary irrigation, toxicity of aluminum and manganese, retention and availability of phosphorus, phosphorus fixation, management and conservation of soils, production systems of rice, soybean, wheat, cassava, corn, and beef cattle.

From 1976/1977 onwards, the Center introduced the practice of central synthesis experimentation in the study of production systems, which aimed to produce one or more representative systems through the use of results from satellite analysis. At the same time, problems detected in central experiments gave rise to new satellite data. Central experiments aimed to identify alternatives to openings and management, recuperation of soil fertility, mechanical practices, integrated pest management, the effects of crop rotation, responses in yields and production, and the economics of systems. Specific methods and techniques of statistical analysis were developed for experimental lots of up to 80 ha. Central experiments also allowed the instigation of experiments at the property level, such as at the Caxuana Farm, in the county of Ponte Nova (MG), where the technology used by the producer was witness to such experiment.

At some point, once obtaining the solutions for certain species, it was necessary to extend the introduction of annual crops in broad network systems, covering the different climatic and soil regions of the Cerrado. Partially as a result of this, in 1981, during the II National Seminar on Soybean Research, new varieties of soybeans were presented and described, showing excellent behavior in the Cerrado, especially Tropical, Doko, and Numbaíra.

In that same year, the Center recommended the cultivation of triticale (a hybrid of wheat and rye) during the rainy season and the dry season, with irrigation. Since the beginning of research into triticale, its productivity had appeared to exceed that of the best cultivars. However, its weight, yield, and quality of flour proved to be inferior.

On the other hand, positive results were obtained with saccharine and grain sorghum, which resulted in indicating their large-scale utilization as the 'little harvest' (*safrinha*) crop, shortly after the main crop, during the rainy season, usually by using short-cycle cultivars.

In 1981, the following perennial crops were being recommended by the CPAC: coffee, fruits (citrus, mangoes, and avocados – germ plasma bank) and forest species, experiments with a mix of *Eucaliptus camaldulensis*, *Pinus oocarpa*, cassava, *Brachiaria spp.* and *Stylosanthes spp.*, as well as forages and pastures – with an active bank of germ plasma with 1,023 species, 900 of which were leguminosae and 123 graminea.

In 1978, the Center achieved its institutional set up with activities in 2,140 ha of experimental and technical-scientific research areas, with a soundly organized team, a well-organized bibliographic collection, 12 operational research laboratories, and full program coverage, upon the inclusion of the agro-forestry-pastoral issue in its activities, actions and tasks.

The rest of that decade served to consolidate the research discoveries and results in the process of development, with intense activity in knowledge and technology transfer, along with technical assistance and rural extension. The CPAC's technical report for 1982–1985, published in 1987, covers the developments in the various research and experimentation fields.

## 2   1985–1994

The results of the work conducted in this period are reported in three technical reports (1985–1987, 1987–1990, and 1991–1995), in addition to publications in other fields such as specialized magazines and JICA–CPAC cooperation reports. The most important event of this period was the VII Symposium on the Cerrado, with the objective of analyzing Cerrado utilization strategies, in April 1989.

As was mentioned, from a model focused on few experimentation and research projects, EMBRAPA decided to choose, from 1981/1982 on, a diffuse model of agricultural research so that each assay, test or experiment would constitute a project assigned to a researcher. Such measure not only made the adoption of a systemic focus of the corporation difficult, but also the coordination of CPAC, both internally and externally with its collaborators. The establishment of the National Research Program (PNP) reduced the difficulties, and the Center started to operate in three PNP's (former projects): (i) National Research Program of Evaluation of Natural and Cerrado's Socio-economic Resources, hereafter PNP

Evaluation; (ii) National Research Program of Utilization of Cerrado's Natural and Socio-economic Resources, hereafter PNP Utilization; and (iii) National Research Program of Production Systems for the Cerrado, hereafter PNP Production Systems.

*(a) PNP evaluation*

The considerable knowledge and information generated between 1976 and 1988 allowed environmental analyzes at the local, regional, and macro-regional levels to be conducted. In the first period (1985–1987), the PNP Evaluation obtained results in the following research fields: natural and introduced ecosystem dynamics; land systems (landscape units) in the Cerrado region; socio-economics and infrastructure. It also developed analytical systems including remote sensing techniques.

Between 1987 and 1990, intense technical training of PNP evaluation researchers was conducted through postgraduate and specialized short courses and specialized technical consultancies, focusing on the following research lines: (i) environmental characterization; (ii) survey and adjustment of existing studies and information; (iii) socio-economic characterization; (iv) macro- and microeconomic analysis of the agricultural industry of the Cerrado; (v) development of an information system with analytical procedures for the storage/processing of geo-environmental and remote sensing information; and (vi) environmental analysis: regional development program, evaluation of germplasm, utilization of native species, environmental impact appraisal, proposition for agricultural use of alternatives, and expansion of the agricultural frontier.

In 1992, through this PNP, actions were undertaken to promote discussion about the new research program model, and to decide on the best way of implementing EMBRAPA's Research Planning System (SEP) in the Cerrado region.

In the annual programming meeting, only one proposal for a new project was submitted and approved. The innovations of the PNP Evaluation only occurred in the development of quantitative method techniques and systems for the improvement of research quality in the analysis process to build demand for the support of small rural productions, in the *Silvânia* Project, analysis of commercialization channels from *Silvânia* county, and the geo-reference databank for this Municipality.

*(b) PNP utilization*

At the beginning of the second decade, research focused on providing solutions to the identified problems, which involved studies on: (i) critical levels, residual effects, and sources of nutrients; (ii) methods of fertilizer application; (iii) recycling of main macro- and micronutrients in the most important annual crops and superficial and sub-superficial altering of soils, with special attention to the use of gypsum; (iv) biological nitrogen fixation (starting in 1993), and reduction or elimination of the use of nitrogen fertilizers in peas; (v) micorrhizae, particularly the pinus species; (vi) the effects of green fertilizers and crop remains; (vii) loss of water and nutrients; (vii) nutritional needs of crops; (ix) physical modification of soils; (x) dynamics of organic matter decomposition; (xi) irrigation, including defining when to commence irrigation, and the sequence of crops; (xii) agricultural machinery and equipment; and (xiii) adjustment and adaptation of machinery for production systems in use.

*(c) PNP production systems*

The CPAC's technical report for the period between 1991 and 1995 presented the results of 48 experiments on soils and water, 51 tests and assays on vegetal production, and 29 assays and experiments on animal production.

The period was fruitful with respect to the production of technical-scientific publications: 516 articles were published and 22 theses were developed by CPAC technicians, 12 at Master's degree level and 10 at Doctor's degree level.

In general, the second decade of the CPAC's existence shows, on the one hand, the continuity of the work carried out in the three national projects (Inventories, Utilization, and Production Systems) and, on the other hand, in the second half of the decade, the strong contribution to knowledge by scientific production and by Master's and Doctor's degree theses.[11]

## 3   1995–2010

In this third period, three events are noteworthy: (i) VIII Symposium on the Cerrado, with a focus on biodiversity and on the sustainable production of food and fibers, in March 1996; (ii) Symposium on Ecology and Biodiversity, in June 2002; and (iii) IX National Symposium on the Cerrado, with the theme: Challenges and strategies for the balance between society, agri-business and natural resources, in 2008.

It is worth mentioning the Center's significant contribution to the Brazilian economy, with technologies applied to crops, such as: (1) rhizobium bacterium, for nitrogen fixation in seeds, when applied to soybeans – positive results are shown, as well as in beans, peanuts, peas and lentils. This technology reduces the costs of nitrogen application, saving approximately US$5.0 billion/year, which makes Brazilian soybean production costs the smallest in the world; (2) direct cultivation is generating savings on fertilizer and machinery costs, equaling approximately R$1.0 billion/year; and 3) genetic improvement, with adaptation of soybean varieties, started in the late 1970s, at the time when the plant grew less than 20 cm tall and productivity was around 1,000 kg/ha, but with good cultivation practices, the average productivity has increased to 3,000 kg/ha.

These three technologies and other practices made it possible to obtain the current level of soybean productivity for commercialization in the Cerrado. As of today, the production of 4,920 kg/ha, and the national average of 3,012 kg/ha, are greater than productivity in the United States and China.

The success of soybeans as the leading crop in development is influencing other segments. For example, Brazil is currently the world's largest producer of sugarcane, confirming Brazilian development in this segment as one of its most successful, being responsible for constant increases in productivity. From an economic standpoint, sugarcane is an expanding sector conditioned by the continuous increase in sugar and alcohol demands and by the revision of the energy matrix toward the use of biomass. It is thus also of key importance to the federal government, which places Brazil as a reference point in the production of sugar, alcohol, and bio-energy.

With a new perspective on agri-business, from the mid-1990s, in addition to soybean and corn development, Brazilian cotton was highlighted again internationally and listed as one of the big export commodities in 2001. Today, the country is the world's fifth largest producer and cotton has one of the best productivity rates of non-irrigated crops.

A study conducted by Nogueira Junior et al. (2002), analyzing the technology and behavior of productivity, in the period between 1980 and 2000, for cotton cultivation in the main producing states – GO, MS, MT, PR, and SP – concluded that the state of GO had the highest frequency of high productivity rates throughout the 1980s, until the mid-1990s. The state of MT went from last place to first regarding productivity, which shows a positive response to the high technological standards adopted in recent harvests. Thus, the Mid-West region increased from 9 percent to 64 percent of total production in Brazil between the tri-annual periods of 1980–1982 and 1998–2000. Between 1975 and 2006, it went from 21 percent to 88 percent. Similarly, the state of BA became the second biggest national producer due to major increases in productivity. According to Barbosa (2002), between 1998–1999 and

2001–2002, its cultivated area increased by 57.8 percent, production by 363.5 percent, and productivity by 165.2 percent, increasing from 950 to 2,520 kg/ha of plume cotton. The three major producing municipalities – São Desidério, Barreiras, and Luís Eduardo Magalhães – were responsible for 72 percent of total state production in 2004, according to IBGE (2005). However, MT is the most outstanding state in this field, contributing 54.36 percent of total cotton production. Whereas in the early 1980s the crop occupied only 4,500 ha of the total cultivated area, which produced a little over 4,900 mt, in the harvest of 2002/2003 it reached an area of approximately 330,000 ha and a record production of more than 500,000 mt.

Unquestionably, the evolution of cotton cultivation in the Mid-West was due to successive increases in productivity. While in the South, represented by the state of PR, the average productivity is 2,388 kg/ha and in the Southeast, 2,448 kg/ha of seed cotton, the Mid-West region obtains an average of 3,550 kg/ha, 47 percent higher.

Briefly speaking, there is programmatic logic in which natural and socio-economic resources were and continue to be evaluated, studied in their relations to soil-climate-plant/animal-agricultural/zoo-technical/forestry practices for the corresponding utilization, according to the characteristics and purposes, and the development of appropriate production systems, in compliance with the agricultural and environmental concepts of sustainable development. In this respect, the research work in the Crop-Livestock-Forest Integration Program (Programa Integração Lavoura-Pecuária-Floresta, ILPF) is gaining the confidence of producers according to the approach of the ABC's Low Carbon Agriculture Program.

In contrast, everything indicates that PNP Utilization and PNP Production Systems were oriented towards the generation of tacit or cognitive knowledge of individual interests, not necessarily towards the interests of organizations and 'caretakers', which can be discussed in a future agenda, for a better reflection on the theme.

## Notes

1. The chapter has been prepared in two stages: an initial and preliminary stage of preparation of its "basic proposal" and work plan; and the second stage that consisted of elaborating on an article regarding the history of development in the Brazilian Cerrado, containing a consultation report and the JICA-CPAC Cooperation.

2. 'The Brazilian scientists dedicating themselves to study the Cerrado are perhaps non-celebrated heroes in the field of savanna ecology, which is quickly developing. The organization of most of these investigations, in a multidisciplinary basis, has increased the savannah's value a lot in the past two decades.' (Prof. Theo L. Hills. McGill University, Montreal, Canada, 1971).

3. 'In 1962, I was a first year student in agronomy, when in a botany class, Prof. Antônio Alvarez de Souza Soares Sobrinho said that the future of grains and cereals production in Brazil will be concentrated in the Cerrado and that Rio Grande do Sul would be reserved for the cultivation of rice and fruits, for the production of beverages and for high-value cattle husbandry. As we see, at that time, the Cerrado was discussed in terms of botany' (Elmar Wagner, 2011).

4. 'Until recently, the Cerrado was considered a useless area for the national economy, and as something undesirable under the aesthetics and even faunistic and forestal aspects' (Paulo Nogueira-Neto, IV Symposium on the Cerrado, Brasília, DF, 1976).

5. Jeca-Tatu is a character created by novelist Monteiro Lobato. He is a poor Brazilian countryman, not assisted by the authorities.

6. *Veranico* is name given to the short dry spell within the rainy season.

7. See the Appendix of this chapter for the full text of 'Technical-scientific production generated and/or validated by the direct action of the CPAC'. In this section, a summary is presented.

8. "The Cerrado needs to be developed because of the population increase and because of the improvement of Brazilians' life conditions, and also to provide supplies to other parts of the world. Japan agreed on cooperating with this development". (Dr. Shiro Okabe, Director)

9. Dr. Sakurai, first Mission/Team Leader, when accompanying Secretary Nair Seiko, was asked by a shop clerk why he could not speak Portuguese. He said that it was because he was Japanese, to which the shop clerk answered, 'A Japanese in Brazil is Brazilian.' So soon after his arrival, this made him and his wife feel welcome.

10. The full list of technical-scientific publications from the JICA–CPAC cooperation is to be found in the Project History Museum of JICA.

11. Soybean producers, researchers, and professors from the United States were surprised to find, in the CPAC's experimental fields, soybeans being seeded, germinating, blooming and being harvested on the same day.

## References

Batistela, M. and E.L. Bolfe (2010) *EMBRAPA satellite monitoring* (Campinas: São Paulo).

Brazilian Association of Technical Standards (1994) *Presentation of articles in periodic publications – NBR 6022* (Rio de Janeiro).

Campos da Rocha, C. M. (2010) *Report of the participation in technical mission to Mozambique*. Brasília/DF, July 11, 2010.

CPAC (2009) *International portfolio of EMBRAPA Cerrados* (Planaltina: EMBRAPA).

EMBRAPA (1975) *Pre-project of implementation of the research center for the development of the Cerrado's resources* (Brasília: EMBRAPA).

EMBRAPA (1976) *Annual technical report of the center for Cerrado agricultural research, 1975–1976. 2nd Edition. Volume 1* (Planaltina: EMBRAPA).

EMBRAPA (1978) *Annual technical report of the center for Cerrado agricultural research, 1976–1977. Volume 2* (Planaltina: EMBRAPA).

EMBRAPA (1979a) *Annual technical report of the center for Cerrado agricultural research, 1977–1978. Volume 3* (Planaltina: EMBRAPA).

EMBRAPA (1979b) *Research results* (Brasília: EMBRAPA).

EMBRAPA (1980) *Annual technical report of the center for Cerrado agricultural research, 1978–1979. Volume 4* (Planaltina: EMBRAPA).

EMBRAPA (1981) *Annual technical report of the center for Cerrado agricultural research, 1979–1980. Volume 5* (Planaltina: EMBRAPA).

EMBRAPA (1982a) *Annual technical report of the center for Cerrado agricultural research, 1980–1981. Volume 6* (Planaltina: EMBRAPA).

EMBRAPA (1982b) *Annual technical report of the center for Cerrado agricultural research, 1981–1982. Volume 7* (Planaltina: EMBRAPA).

EMBRAPA (1987) *Annual Technical Report of the Center of Cerrado's Agricultural Research. 1982–1985* (Planaltina: EMBRAPA).

EMBRAPA (1991) *Annual technical report of the center for Cerrado agricultural research, 1985–1987* (Planaltina: EMBRAPA).

EMBRAPA (1994) *Annual technical report of the center for Cerrado agricultural research, 1987–1990* (Planaltina: EMBRAPA).

EMBRAPA (1997) *Annual technical report of the center for Cerrado agricultural research, 1991–1995* (Planaltina: EMBRAPA).

EMBRAPA (2010) *Pro-savannas* (Brasília: EMBRAPA).

EMBRAPA/CPAC (1975) *Project of implementation of the center for Cerrado agricultural research* (Brasília: EMBRAPA).

EMBRAPA/CPAC (1976) *Program of the center for CPAC agricultural research – 1976/77* (Planaltina: EMBRAPA).

EMBRAPA/CPAC (1977) *IV Symposium on the Cerrado. July 1976. Volume I* (Brasília: EMBRAPA).

EMBRAPA/CPAC (1979a) *Research results obtained in CPAC in 1979* (Planaltina: EMBRAPA).

EMBRAPA/CPAC (1979b) *V Symposium on the Cerrado: Use and management* (Brasília: EMBRAPA).

EMBRAPA/CPAC (1988) *VI Symposium on the Cerrado. Savannas: Food and energy* (Brasília: EMBRAPA).

EMBRAPA/CPAC (1995) *VII Symposium on the Cerrado: Strategy of utilization.* (Brasília: EMBRAPA).

EMBRAPA/CPAC (1996) *VIII Symposium on the Cerrado and 1st International Symposium on Tropical Savannas*, March 24–29 (Brasília: EMBRAPA).

EMBRAPA/CPAC (2008) *IX National Symposium on the Cerrado – Savannas: Challenges and strategies for the balance between society, agribusiness and natural resources, and 2nd International Symposium on Tropical Savannas* (Planaltina: EMBRAPA).

EMBRAPA/CPAC (2009) *Savannas: Requests for research. 1st Edition* (Brasília: EMBRAPA).

EMBRAPA/DID (1981) *Cerrado: Information abstracts. Volume 3* (Brasília: EMBRAPA).

Gelape Faleiro, F. and A. Lopes de Farias Neto (eds) (2009) *Savannas: Requests for research.* (Planaltina: EMBRAPA).

Goedert, J.W., E. Wagner, and A. Barcelos (2008) *Tropical savannas: Dimensions, history and perspectives*. In: *IX National Symposium on the Cerrado and International Symposium on Tropical Savannas, Annals* (Brasília).

Goodland, R. (1979) 'History of works in the Cerrado until 1968' in R. Goodland and M.G. Ferri, eds, *Ecology of Cerrado* (São Paulo: USP; Itatiaia: Belo Horizonte), pp. 13–21.

IPEA (1973) *Current and potential use of the CPAC: Physical basis and potentialities of the region* (Brasília: IPEA. Series studies for planning, 2).

JICA/CPAC (1980) *Cooperation agreement in agricultural research in Brazil entered into on 30/09/1977. 1st Team: Partial report of the project of cooperation in agricultural research in Brazil, 1978–1980* (Brasília: JICA/CPAC).

JICA/CPAC (1986a) *Cooperation agreement in agricultural research in Brazil entered into on 30/09/1977. 2nd Team: Technical-scientific works developed by the cooperation project in agricultural research in the Brazil, 1980–1983* (Brasília: JICA/CPAC).

JICA/CPAC (1986b) *JICA–CPAC cooperation agreement. Results of the Japanese–Brazilian cooperation in agricultural research. 3rd Team: Considerations on the agricultural development of the Cerrado* (Brasília: JICA/CPAC).

JICA/CPAC (1994) *Cooperation Agreement in Agricultural Research in Brazil entered into on 30/08/1987. 1st Team 1987–1990. 2nd Team 1990–1992. Technical Report of the Japanese–Brazilian Cooperation in Agricultural Research in the Cerrado. 1987–1992* (Brasília: DF).

Kubota, Toru (1966) *Scientific contribution of the JICA project to sustainable agricultural development in the Cerrado. VIII Symposium on the Cerrado and 1st International Symposium on Tropical Savannas* (Brasília).

Lemos, A.A.B. (ed.) (1976) *Cerrado: Analytical bibliography* (Brasília: EMBRAPA, Department of Information and Documentation).

Müller, Mary Stela and Julce Mary Cornelsen (1994) *Rules and standards for theses, dissertations and monographs* (Londrina).

Rosavannas (2010) *Project for improvement of research and technology transfer capability for the Development of Nacala Corridor agriculture, in Mozambique* (Draft).

Various publications on the website http://www.ainfoweb.cnptia.embrapa.br.

# 8
# Environment-Friendly Land Use of the Cerrado

*Edson Eyji Sano*

## Introduction

The Cerrado has been named as one of the world's 25 priority areas for biodiversity conservation. The headwaters of three major basins of Brazil are located in the Cerrado: Paraná, Araguaia/Tocantins, and São Francisco. There are over 12,300 different plant species. Many of them have yet to be evaluated scientifically and economically and may have high potential for medicine and cosmetics. This chapter aims to appraise how environment-friendly land use has been attained in the Cerrado from the standpoint of ecological and environmental conservation. First, an overview of current land use by agriculture and livestock activities in the Cerrado is presented (Section 8.1). Then, the Cerrado's land use is analyzed in detail using Landsat satellite images (Section 8.2). Bearing this information in mind, public policies aimed at environment-friendly land use of the Cerrado are discussed in Section 8.3, while in Section 8.4, the contribution by farmers to such land use is analyzed. Finally, in Section 8.5, some of the major future challenges for environmental and ecological conservation in the Cerrado are discussed.

## 8.1  Overview of land use by agriculture in the Cerrado

The Brazilian tropical savannah (nationally known as the Cerrado) occupies the central part of the country, covering 204.7 million ha, about a quarter of Brazil. It partially covers the following states of Brazil (IBGE, 2004): Bahia (BA), Goiás (GO), Maranhão (MA), Mato Grosso (MT), Mato Grosso do Sul (MS), Minas Gerais (MG), Paraná (PR), Piauí (PI), São Paulo (SP), and Tocantins (TO), and the Federal District (DF) (see Figure 8.1). The Cerrado extends from the coastline of MA to the southern states of

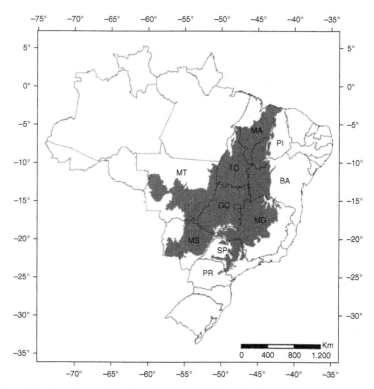

*Figure 8.1* Location of the Cerrado biome in Brazil. Capital letters indicate the states covered by this biome. Bahia (BA), Goiás (GO), Maranhão (MA), Mato Grosso (MT), Mato Grosso do Sul (MS), Minas Gerais (MG), Paraná (PR), Piauí (PI), São Paulo (SP), and Tocantins (TO), and the Federal District (DF)

*Source:* IBGE (2004).

SP and PR – a variation of 22.4° in latitude and an elevation range from sea level to 1,800 meters. This highly heterogeneous biome in terms of biodiversity has strong seasonality – wet from October to March and dry from April to September (Ferreira and Huete, 2004). The Cerrado's natural vegetation consists of mixtures of grassland – often with sparse occurrences of shrubs and small trees – shrubland, and woodland (Ribeiro and Walter, 2008). Riparian forests are found along watercourses.

At this point, it is worthwhile introducing a comment about the size and boundaries of the Cerrado. Some authors consider the size of the Cerrado as 203.7 million hectares, others as 207.4 million hectares.[1] The differences come from the difficulty in specifying the exact boundary

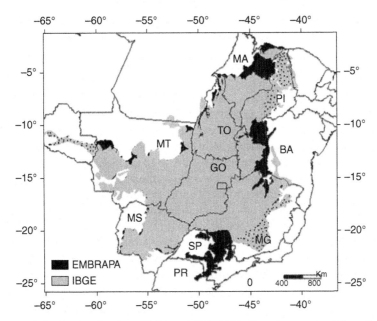

*Figure 8.2*  Overlay between IBGE and EMBRAPA Cerrados maps of Cerrado's boundaries

*Notes:* (1) Black zones are areas that are not included in EMBRAPA's definition. Dotted zones are areas that are not included in IBGE's definition; (2) This map has been prepared for illustrative purposes. For a precise definition of the Cerrado by EMBRAPA, see the map in the Introduction chapter of this book. For a precise definition by IBGE, see Figure 8.1 above.

of the Cerrado because the transition to other ecosystems is gradual. In 2002, the Brazilian Institute for Geography and Statistics (IBGE) published the official map of the biomes of Brazil at a scale of 1:5,000,000 (IBGE, 2004). This map is the one used throughout this chapter, according to which the area of the Cerrado is considered to be 204.7 million ha. Figure 8.2 shows the major differences between the official IBGE map and the previous, unpublished, and broadly used map prepared by CPAC (EMBRAPA Cerrados). Basically, new areas were included in the southern (SP), western (BA), and northern (south of PI) parts of the Cerrado; in addition, areas of Rondonia (RO) State (not shown in the figure) were excluded.

Until the middle of the 1960s, the sparse land use in the Cerrado was characterized by environment-friendly cattle ranching over natural grasslands. Sometimes, adventurers (*bandeirantes*) also ventured into this area. They came from the southeast of the country, looking for precious minerals, especially gold and emeralds (Brannstrom et al., 2008). In the Cerrado, only hardy and siliceous shrubs and trees could survive.

Then, research arrived. An agriculture research center was created by the federal government in Planaltina, a satellite city 35 km northeast of Brasília: 'EMBRAPA Cerrados', formerly the Cerrado Agricultural Research Center (CPAC), belonging to the Brazilian Agricultural Research Corporation (EMBRAPA), under the Ministry of Agriculture Livestock and Supply (MAPA). After rigorous soil laboratory analyzes and field experiments at the CPAC, recommendations were introduced for the addition of limes and fertilizers to more appropriately attend to the needs of the crops. The major issue was to determine the correct amount of nutrients to replace what nature had taken out from the upper soil layers of the Cerrado during the geological leaching process. Between 4 and 5 mt of lime per hectare started to be scattered over the 30 to 50 surface layers to correct soil acidity, a technique known as liming. Farmers came from the traditional agricultural areas of the southern (*gauchos*) and southeastern (*paranaenses*) part of Brazil, searching for new opportunities and cheap land.

New varieties of pastures from Africa were developed. They were named *Braquiarinha* (*Brachiaria decumbens*) and *Braquiarão* (*Brachiaria brizantha*) and they were more biomass productive and more adapted to acid soils. The development of biological nitrogen fixation by a bacterium named *Rhizobium*, a study led by a former EMBRAPA researcher, Dr. Johanna Dobereiner, has been helping the country to save billions of dollars in chemical fertilizers for soybean production.

*Chapada* (extensive flat terrains), which occur in several parts of the Cerrado, are a very attractive landscape pattern for intensive crop production. Flat topography facilitates mechanized operations for crop production. In addition, such regions often have enough precipitation for rain-fed grain production (i.e., without irrigation). Roughly speaking, the threshold for rain-fed agriculture in the Cerrado is 1,000 mm of rainfall per year.

The six-month dry season has its benefits, however. First, it helps to decrease the diseases commonly found in the more humid regions of Brazil (south and southeast). This is the case, for instance, in the occurrence of aflatoxins or ochratoxins in coffee and other grains produced in the Cerrado, a relevant aspect for food exportation. Second, it means more solar radiation, improving the flavor of coffee and fruits.

However, the availability of farmland suitable for agriculture was only a secondary reason for the extraordinary growth of agriculture in this region. The primary reason was the EMBRAPA and Japan International Cooperation Agency (JICA) agricultural development cooperation project, which created a model for agricultural development in tropical areas throughout the world.

The Cerrado is now a major food producer in Brazil. In 2002, about 54 million ha of cultivated pastures and 21 million ha of crops were to be found in this biome (Sano et al., 2010). There are four well known regions characterized by large-scale, high-tech grain production, mainly soybeans, maize, and cotton. The regions are: Luís Eduardo Magalhães (in the west of BA), Lucas do Rio Verde (MT), Rio Verde (southwest of GO), and Balsas (south of MA). Parts of these regions have already turned into 'white oceans of cotton,' according to an article entitled 'The Miracle of Cerrado,' published in *The Economist* magazine on August 26, 2010. These regions have the highest soybean productivity in Brazil, at over 3,000 kg per hectare. To the south of MA and PI is the newest agricultural frontier of the Cerrado. The physical properties of the soil in this region are better than those found, for instance, in MT and BA; land is still cheap; and the distance to Itaqui port in São Luís, the capital city of MA, for grain exportation, is not so great.

## 8.2 Cerrado land use as seen from space

An analysis of 170 Landsat satellite images obtained in 2002 indicated that about 40 percent of the Cerrado (80 million ha) were used mostly for grain and meat production (Sano et al., 2010) (Table 8.1). This corresponds to about 55 percent of the total amount of land suitable for intensive agriculture in this biome, which is approximately 135 million ha (estimation based on soil properties and topography).

The spatial distribution of land use is highly heterogeneous. The majority of the land use in Cerrado occurs in its southernmost part

*Table 8.1*  Approximate area occupied by different land use classes in the Cerrado biome, estimated from the analysis of Landsat satellite images (2002)

| Classes | Area (x 1,000 ha) | Percentage (%) |
|---|---|---|
| Cropland | 21,590 | |
| Pastureland | 54,150 | |
| Urban areas | 890 | |
| Reforestation | 3,170 | |
| SUBTOTAL | 79,800 | 39 |
| Natural Cerrado | 124,000 | 61 |
| TOTAL | 204,700 | 100 |

*Source:* Sano et al. (2010).

(Figure 8.3). On the other hand, the largest portion of original vegetation remaining is found in the northern part. SP, PR, and MS are the states that present the highest indices of land use classes (Table 8.2): 87 percent, 68 percent, and 68 percent, respectively. The three states from the northern part of the Cerrado, that is, PI, MA, and TO, presented the lowest indices of land use: 8 percent, 11 percent, and 21 percent, respectively.

This spatial pattern of land use represents the national history of land occupation itself, which started after the 1920s, when coffee production was the major economic activity in Brazil, mainly in SP. As the land use in SP intensified, use of the southern part of GO was initiated, based upon government subsidies and technical assistance to producers. The northern part of the Cerrado, a region with difficult access and few large urban areas, is still well reserved.

Pastureland (Figure 8.3a) is found mainly in the states of GO (13 million ha), MG (11.8 million ha), and MS (11 million ha). In MS, the majority of land use corresponds to cultivated pasture (mainly *Brachiaria spp.*). In this state, the largest beef cattle herd in Brazil, a relatively good infrastructure, and large numbers of meat processing plants are found. In MG, cultivated pasture is the dominant landscape in a region known as *Triangulo Mineiro*, where 25 percent of the original vegetation is still preserved. In GO, along with cultivated and native pastures throughout

*Table 8.2*   Distribution of land use classes found in the Cerrado per state (2002)

| State | Percentage of state within the Cerrado (%) | Cropland (x 1,000 ha) | Pastureland (x 1,000 ha) | Reforestation (x 1,000 ha) | Total (%) [*] |
|---|---|---|---|---|---|
| SP | 33 | 3,586 | 2,622 | 533 | 87 |
| PR | 2 | 84 | 104 | 67 | 68 |
| MS | 61 | 2,712 | 10,948 | 1,018 | 68 |
| DF | 100 | 137 | 120 | 4 | 56 |
| GO | 97 | 5,038 | 12,932 | 51 | 56 |
| MG | 57 | 2,122 | 11,838 | 1,302 | 47 |
| MT | 40 | 5,561 | 6,509 | 32 | 34 |
| BA | 27 | 1,573 | 2,257 | 126 | 26 |
| TO | 92 | 176 | 4,253 | 377 | 21 |
| MA | 65 | 356 | 1,902 | 27 | 11 |
| PI | 37 | 215 | 522 | 1 | 9 |

*Note:* See Figure 8.1 for state identification; * Percentage of land use in the part of the state covered by the Cerrado.

*Source:* Sano et al. (2010).

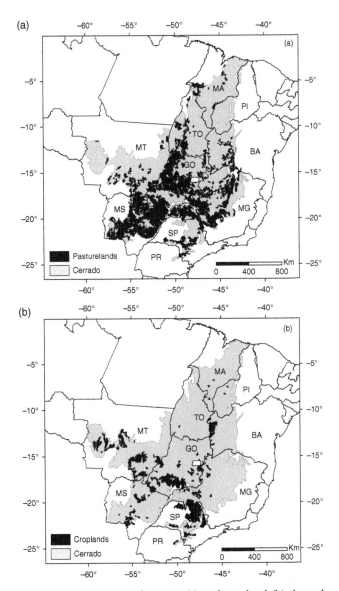

*Figure 8.3* Spatial distribution of pasture (a) and cropland (b) throughout the Cerrado. For state identification, see Figure 8.1

*Source:* Sano et al. (2010).

the state, vast croplands, restricted to the southwestern area, occur near the counties of Rio Verde and Jatai. There, soybeans, corn, and, more recently, cotton and sugarcane are produced with intensive use of machinery. In general, cropland (Figure 8.3b) is found mostly in MT (5.6 million ha), GO (5 million ha), and SP (3.6 million ha).

Estimates of deforestation in the literature show a decreasing trend for the Cerrado since 2008 (IBAMA, 2013). Until that year, more than 14,000 km² of the Cerrado had been deforested; however, after that year, deforestation decreased to levels below 7,700 km² (Figure 8.4). Although the methodology, data set, and time period are different for each evaluation shown in this last figure, there is a clear indication that, for several reasons, farmers are no longer so eager to clear Cerrado vegetation. It is also noticeable that the recent areas of major deforestation are in BA,

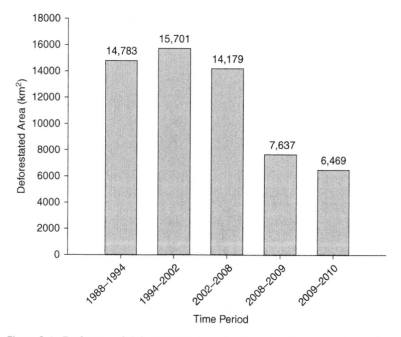

*Figure 8.4* Evolution of deforestation over the Cerrado, showing a decreasing trend after 2002

*Note:* The Ministry of Science and Technology coordinated the estimation for 1988–1994 (Brazilian inventory of greenhouse gases emission from agriculture). The National Space Research Institute (INPE) coordinated the estimation for 1994–2002. The Ministry of Environment and Brazilian Institute for Environment and Renewable Natural Resources (IBAMA) coordinated the other estimations.*Source:* IBAMA (2013).

*Table 8.3* The ten counties of the Cerrado with the highest levels of deforestation in 2008–2009 and 12 counties with highest levels in 2009–2010

| County | State | County area (km²) | Deforested area (km²) |
|---|---|---|---|
| **2008–2009** | | | |
| Formosa do Rio Preto | BA | 16,186.42 | 197.17 |
| Baixa Grande do Ribeiro | PI | 7,808.10 | 168.07 |
| Jaborandi | BA | 9,476.13 | 130.55 |
| Correntina | BA | 12,145.81 | 126.14 |
| São Desidério | BA | 14,820.57 | 123.6 |
| Codó | MA | 4,360.14 | 121.59 |
| Ribeiro Gonçalves | PI | 3,978.15 | 120.54 |
| Barreiras | BA | 7,897.49 | 110.84 |
| Paranátinga | MT | 16,533.11 | 107.04 |
| Grajaú | MA | 7,031.23 | 99.96 |
| **2009–2010** | | | |
| Baixa Grande do Ribeiro | PI | 7,808.83 | 394.29 |
| Uruçuí | PI | 8,453.63 | 203.48 |
| Formosa do Rio Preto | BA | 16,186.06 | 143.92 |
| São Desidério | BA | 14,821.67 | 119.85 |
| Mateiros | TO | 9,593.24 | 93.06 |
| Barreiras | BA | 7,897.58 | 88.39 |
| Balsas | MA | 13,144.33 | 85.24 |
| Santa Quitéria do Maranhão | MA | 1,918.14 | 73.88 |
| Codó | MA | 4,363.32 | 69.91 |
| Riachão das Neves | BA | 5,837.45 | 68.81 |
| Baixa Grande do Ribeiro | PI | 7,808.83 | 394.29 |
| Uruçuí | PI | 8,453.63 | 203.48 |

*Source:* IBAMA (2013).

PI, and MA, in the northeastern part of Cerrado. As shown in Table 8.3, among the ten counties with the highest levels of deforestation in 2008–2009, nine were in these three states. The same applied in 2009–2010.

## 3 Actions for environment-friendly land use of Cerrado: public policies

As was mentioned, the Cerrado has been included as one of the world's 25 priority areas for biodiversity conservation. The headwaters of three major basins of Brazil are located in Cerrado: Paraná, Araguaia/Tocantins, and São Francisco. There are over 12,300 different plant species, according to the book edited by Sano et al. (2008). Many of them have yet to be evaluated scientifically and economically and may have high potential for medicine and cosmetics. The Cerrado hosts *Caryocar brasiliense*, the scientific name of the *pequi*, a tree protected by law and whose fruit is very much enjoyed by local people.

Using these arguments, a number of people support the idea that the Cerrado has been completely destroyed or will disappear in the near future. The 2002 land use map showed that 61 percent of the Cerrado was still covered by natural vegetation. There is no other region in the world with such an intensive food production system preserving such a high-level of natural vegetation. Of course, this percentage does not mean authorization for more illegal deforestation. In fact, the intensity and type of land use reported has caused concern not only for environment-oriented people but also for farmers who want to move to environmentally sustainable agriculture. Soil loss by wind and rainfall erosion, contamination of surface and ground water resources, and high emission levels of greenhouse gases are among the major issues that have been highlighted. Bearing these aspects in mind, public policies aimed at environment-friendly land use of Cerrado are discussed in this section, while in the next section, contributions by farmers in such land use will be analyzed.

The Brazilian federal government is making some contributions for the country to achieve environmentally sustainable agriculture – not only in the Cerrado but also in five other biomes: Amazonia, Atlantic Forest, Caatinga (semi-arid), Pampa (southern natural grasslands), and Pantanal (swamp area). Specifically for the Cerrado, on September 15, 2010, PPCerrado was created, a short name for the Plan for the Prevention and Control of Deforestation and Forest Fires in the Cerrado (MMA, 2013a). The goal of this plan is to reduce the amount of deforestation forecast for 2020 by 40 percent, in addition to creating new conservation units and producing macro-ecological and economic zoning maps. The priority sites of PPCerrado include the major remaining natural vegetation sites of the Cerrado, especially the 20 counties of the Cerrado with the highest levels of deforestation between 2002 and 2008 (Figure 8.5) and those with:

- Intense land use pressure;
- High priority for biodiversity conservation;
- High relevance for conservation of water resources, especially the headwaters of major hydrographic basins.

Monitoring and control, the protection of priority areas for conservation, territorial planning (zoning), and the promotion of environmentally

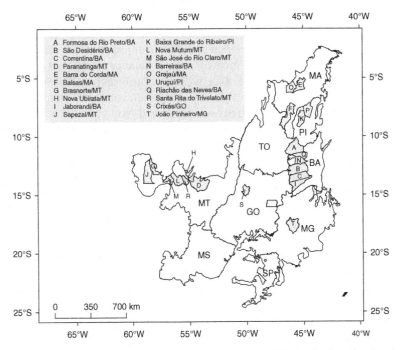

*Figure 8.5* Location of 20 counties of the Cerrado with the highest levels of deforestation between 2002 and 2008

*Note:* See Figure 8.1 for state identification.*Source:* IBAMA (2013).

sustainable activities are the ultimate goals of PPCerrado. Actions to increase protected lands by law include: increasing the area of federal conservation units to 2.5 million ha; conducting ecological and economic zoning of the Cerrado and supporting state initiatives to carry out local zoning; ratifying 300,000 ha of indigenous land; and increasing indigenous land to 5.5 million ha. The promotion of environmentally sustainable activities includes credits to recover 8 million ha of degraded pastures, legal reserves and permanent protection areas as well as credits for reforestation and implementation of agroforestry systems.

Recently, two very important agricultural zonings at national level were promoted by MAPA: the agricultural zoning of climatic risk (Assad et al., 2008) and sugarcane zoning (Manzatto et al., 2009). The agricultural zoning intends to reduce climatic risks related to the production of different rain-fed crops throughout the entire country. It involves the production of state-based maps showing all counties having favorable conditions to cultivate a specific crop at a specific time of the year. The study takes into consideration historical data over 20 years on precipitation, soil type, and crop coefficients (Kc) of different crops. These are also used as the input parameters for a water balance model that defines the planting conditions. The objective is to minimize the risks related to climatic phenomena and to allow each Brazilian county to identify the best planting dates for different soil types and crop varieties. It is very easy to understand and has been adopted by farmers, financial agents, and other users.

This study considers a minimum of 20 years of daily precipitation data and three types of soil texture: clayey soils (>35 percent clay); sandy soils (>70 percent sand and <15 percent clay); and medium texture (a Brazilian terminology for soils with similar proportions of clay, silt and sand). Crops considered in this study include bean, coffee, cotton, maize, millet, sorghum, soybeans, and wheat. Currently, the study is available for 15 annual crops and 24 perennial crops and maize + *Brachiaria* consortium. Each crop is divided into three groups: short, intermediate, and long cycle.

Agricultural zoning in Brazil was first adopted in 1996 for wheat. Since then, it has been reviewed every year and can be accessed on the homepage of MAPA (http://www.agritempo.gov.br). Farmers can access this homepage to find the planting conditions of a given crop in their county. Figure 8.6 shows an example of planting conditions for short-cycle soybeans in clayey soil in the period December 1–10 for counties in GO. As can be seen, all counties except Formosa and São João d'Aliança had favorable conditions for planting short-cycle soybeans in that period. Favorable conditions mean that there will be enough rainfall on the planting date, low probability of the occurrence of a 15-day dry spell (or frost, depending on the study area), and no rainfall during harvesting time.

The study is used by Banco do Brasil, the public institution that provides farm insurance to producers, which means that if farmers do not follow the published recommendations of crop zoning in terms of planting dates, they will not be eligible to receive insurance. Unofficial estimates indicate that the Banco do Brasil was paying out around US$150 million in insurance claims before agricultural zoning was introduced; this total went down to about US$500,000 after the bank started to impose this restriction.

A prime example of environmentally sustainable agricultural practice in Brazil is the production of ethanol from sugarcane. Brazil is one of the few countries in the world with a large number of vehicles that run on

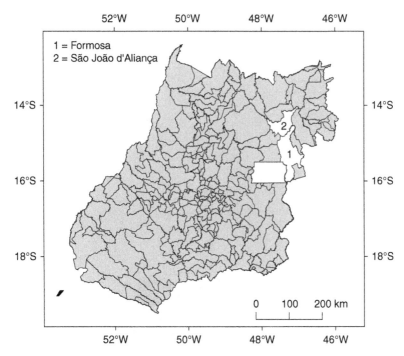

*Figure 8.6* Counties of Goiás state with conditions favorable to planting short-cycle soybeans in clayey soils in the December 1–10 period

*Source:* Ministry of Agriculture/AGRITEMPO.

ethanol. Nowadays, most new Brazilian cars are *flex*, which means that drivers have the option to mix gasoline with ethanol at any proportion or even use only ethanol whenever prices are more attractive. Ethanol is environmentally sustainable because it releases less carbon dioxide into the atmosphere than gasoline. According to the Union of Sugarcane Agroindustry (UNICA) of SP, in 2003, the emission of 27.5 million mt of $CO_2$ equivalent into the atmosphere was avoided by the use of ethanol in place of gasoline.

Figure 8.7 shows the distribution of sugarcane in the Cerrado for 2010. The crop is mostly concentrated in SP (2.5 million ha) and the south of MG (585,000 ha), which is about 76 percent of the total sugarcane cultivation area found in this biome (4,144,000 ha). The south of MS and the south of GO also have significant areas of sugarcane fields.

There is plenty of land that can still be used for sugarcane cultivation. However, the potential expansion of sugarcane cultivation in the

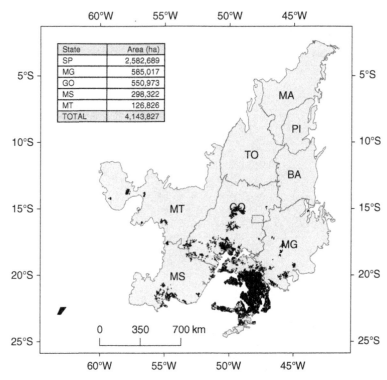

| State | Area (ha) |
|-------|-----------|
| SP | 2,582,689 |
| MG | 585,017 |
| GO | 550,973 |
| MS | 298,322 |
| MT | 126,826 |
| TOTAL | 4,143,827 |

*Figure 8.7*  Distribution of sugarcane in the Cerrado (2010)
*Source:* CANASAT/INPE.

Cerrado to supply internal and foreign demands has raised concerns about the sustainability of the biome. The advance of sugarcane fields in the Cerrado is mostly at the expense of crop fields such as soybeans and maize, because the soils are already corrected for fertility and acidity. Additional costs need to be taken into consideration by farmers whenever sugarcane substitutes for natural vegetation or degraded pasture. Of particular concern is the demand for water. The total amount of rainfall required by sugarcane is estimated at 1,500–2,500 mm per year, which should be uniformly distributed along the growing cycle. This level is problematic for the Cerrado, where mean annual precipitation is around 1,500 mm and there is a well defined six-month dry season. Thus, farmers need to use irrigation, mainly when plants start re-growing after harvesting. The agro-ecological zoning of sugarcane was therefore launched by MAPA in 2009. The objective was to provide technical information for decision makers to formulate public policies for the sustainable expansion of sugarcane in Brazil. The following criteria were adopted in this study:

* Amazônia and Pantanal, the two most environmentally sensitive biomes of Brazil, were excluded;
* Natural vegetation-covered areas and protected areas by law (e.g., conservation units and indigenous lands) were also excluded;
* Land with slopes of >12 percent (unsuitable for machine harvesting), rocky and sandy soils, insufficient rainfall (water deficit >400 mm), high temperatures (average annual temperature < 19°C) were also excluded.

Estimates have demonstrated that the country has 64.7 million ha of land suitable for sugarcane expansion. Considering only the land occupied by cultivated pasture, 37.2 million ha is still available. This means that Brazil does not need to incorporate any natural vegetation-covered land for renewable energy production or reduce land used for grain production. Regarding the states covered by the Cerrado, the major land for sugarcane expansion is located in GO (Table 8.4). In this state, there are more than 12 million ha of potential land for sugarcane cultivation, the majority cultivated pasture land.

The Rural Environment Register (CAR) is another important initiative of the federal government in terms of preserving the Cerrado's environment (MMA, 2013b). First of all, farmers provide the geographical coordinates of the limits of their land for the Geographical Information System (GIS)-based, CAR system. Next, authorized technicians map and

*Table 8.4* Land (in hectares) suitable for sugarcane expansion in GO, MT, and MS, the major states covered by the Cerrado biome (2002) (TO in the northern part of the Cerrado was excluded in the agro-ecological sugarcane zoning because this state is considered part of Legal Amazônia.)

| State | Land use | | |
|---|---|---|---|
| | Cultivated pasture | Agriculture | Total |
| Goias | 7,782,000 | 4,812,000 | 12,594,000 |
| Mato Grosso | 2,582,000 | 4,231,000 | 6,813,000 |
| Mato Grosso do Sul | 6,247,000 | 2,458,000 | 8,705,000 |
| TOTAL | 16,611,000 | 11,501,000 | 28,112,000 |

*Source:* Manzatto et al. (2009).

monitor the land use and land cover changes based on satellite images. Farmers registered in the CAR and practicing environmentally correct use of their land receive different types of advantages, for instance, reduction of fines or punishments. Therefore, CAR is:

- a prerequisite for obtaining environmental licenses for any economic activities on a farm;
- evidence of the commitment of producers to comply with their environmental obligations; and
- one of the best ways for farmers to respond to the pressures of society and markets anxious about the preservation of the environment in food production activities.

## 8.4   Actions for environment-friendly land use of the Cerrado: contributions from farms

Environmentally sustainable agricultural technologies and crop management developed by research will only be adopted by producers if they reduce production costs or if they bring in extra revenue. This is the case of no-till farming, which has been widely adopted by Cerrado farmers (Bolliger et al., 2006; Sano and Pinhati, 2009). This is a farming system that eliminates all plowing and most other tillage operations, increasing residual soil cover and reducing costs for producers (less fuel and labor and prolonged equipment life). After harvest, the crop residues are left on the soil surface as mulch and the following crop is planted directly into the remaining residue. The soil surface is rarely left bare, so that erosion by wind or water runoff is greatly reduced and soil moisture is kept higher. The continued addition of organic matter from crop residues improves the soil structure. Ultimately, these all combine to reduce, for example, water pollution, since less sediment and chemicals are removed from the fields. No-till farming systems often include some green cover plants (e.g., millets and legumes), which are planted right after crop harvesting. They are a source of nitrogen for crops. The drawbacks are that special machinery is required to sow in the residues, and additional chemicals may be needed to better control weeds. Diseases and insects that damage crops may also occur more intensely than with conventional tillage systems.

In a study conducted by Bastos Filho et al. (2007), 97 percent of producers from GO, TO, western BA, southeastern PI, and southern MA (Figure 8.8) said that they were adopting a no-till system partially or for the entire farm. The two major reasons for adopting a no-till system,

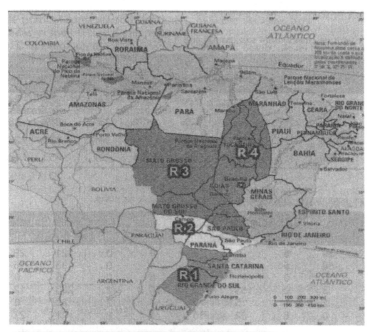

| ADOPTION | R1 | R2 | R3 | R4 |
|----------|-----|-----|-----|-----|
| Yes | 100 | 97 | 95 | 93 |
| No | 0 | 3 | 5 | 7 |

| Reasons | R1 | R2 | R3 | R4 |
|---------|-----|-----|-----|-----|
| Productivity | 79% | 59% | 63% | 53% |
| Soil conservation | 92% | 89% | 94% | 95% |
| Costs | 19% | 22% | 20% | 20% |
| Not especified | 3% | 9% | 7% | 5% |
| Number of answers | 154 | 64 | 227 | 40 |

*Figure 8.8*   Field research (questionnaire) showing the level of adoption of a no-till system by farmers from different regions of Brazil

*Note:* The R4 region encompasses the Cerrado.*Source:* Bastos Filho et al. (2007).

they pointed out, were an increase of soil conservation and an increase in productivity. In conventional tillage systems, soil loss by erosion can reach 25 mt/ha/yr. In no-till systems, soil loss can be reduced to about 3 mt/ha/yr.

In another field survey conducted by the author in November 2006, the author visited 98 fields of soybean, maize and cotton located in Luís Eduardo Magalhães County. (western BA) (Sano and Pinhati, 2009) Seventy three farms were using a no-till system, that is, 74 percent of the farms visited in the survey. Another drawback of the no-till system in the central part of Brazil is the speed of natural degradation of dry biomass left on the soil surface, which is much greater than in the south of Brazil. Bastos Filho and his collaborators estimated that the coverage of crop residue on the soil surface in the central part of Brazil was between 25 and 35 percent, while the percentage for the southern part of the country ranged from 42 to 57 percent. Dry biomass from crops like soybeans decomposes quickly because of the low C:N ratio and low amounts of lignin and cellulose.

Crop–livestock integration (Trecenti, 2011) is another technique that has had high acceptance among farmers. It is particularly suitable for animal producers facing pasture degradation. In the Cerrado, it is estimated that 60–70 percent of cultivated pasture has some level of degradation. As a consequence, the Cerrado's cultivated pasture can host only 0.5 head of cattle per ha on average. Crop–livestock integration can be used as a type of pasture management to recover degradation within four years. The following is the action taken by a farm, where 100 percent of its land was degraded pasture, completely recovered after four years:

Year 1:   Cattle were removed from a quarter of the farm and soybeans were planted (with liming and fertilization);

Year 2:   Another quarter of the farm was converted to soybeans or other crops, while pasture was re-planted in the first quarter of the farm;

Year 3:   The same procedure as in Year 2 for another quarter of the farm;

Year 4:   The same procedure as in Year 2 for the last quarter of the farm.

At the end of the fourth year, all the pasture was back to cultivated pasture, and the farmer had obtained extra revenue from crop production. However, this type of management works better if the farmer is experienced in both animal and vegetal production or in places where grains can be sold easily. Ranchers can also rent the farmer's land for experienced grain producers.

There are several benefits of crop–livestock integration, such as:

• Increase in milk and meat production from higher biomass productivity of pasture;
• Increase in grain production from additional areas of cropland;
• Reduction of pasture degradation;
• Reduction of soil erosion and water contamination;
• Reduction of pressure for new deforestation;
• Diversification of products at farm level.

More recently, reforestation was included in crop–livestock integration so that there are now farmers with crop–livestock–forestry integration: all three together or any two. Eucalyptus provides shade to cows and to the pasture beneath, which remains green for longer during the dry season. After three to four years, depending on the variety, it can be harvested and sold as timber for fencing or firewood or as a source of cellulose, among other uses.

In western BA (more specifically in the counties of Luís Eduardo Magalhães and Barreiras), several areas were using the center-pivot irrigation system for coffee production. Since irrigation was done on an almost daily basis, without any regard for crop water necessity, there was clearly excessive water use, with consequent additional costs in energy and the risk of surface and ground water contamination.

Dr. Euzebio M. da Silva, a former researcher at CPAC (EMBRAPA Cerrados), proposed a project to reduce water usage by installing tensiometers (portable moisture sensors) in irrigated areas. The study site was the Agribahia Farm (county of Luis Eduardo Magalhaes, BA; −11.86° south, −45.73°west), a farm with six irrigation systems for coffee production (330 ha with a center-pivot irrigation system and 160 ha with a dripping irrigation system).

Tensiometers were installed in the two fields irrigated with center-pivots, and soil moisture content was measured from July to October: Pivot # 1 was irrigated according to the measurements of the tensiometers; and Pivot # 2 was irrigated daily at the maximum capacity of the pivot (the irrigation procedure then adopted by local coffee growers).

The results are shown in Table 8.5. In July, the producer was applying 50.5 percent more water than the plants actually needed. However, the difference decreased to 3.8 percent in October because of the increase in evapotranspiration. This study demonstrated that it is possible to reduce

*Table 8.5* Comparison of monthly amount of water applied in Pivot # 1 (controlled by tensiometer) and in Pivot # 2 (daily irrigation at maximum capacity of equipment) in Agribahia Farm during the dry season of 2004

| Month | Reference evapotranspiration (ET, mm) | Amount of applied water (mm) | | Reduction of water supply (%) |
|---|---|---|---|---|
| | | Pivot # 1 | Pivot # 2 | |
| July | 127.2 | 104.3 | 210.8 | 50.5 |
| August | 158.4 | 134.1 | 202.7 | 33.8 |
| September | 186.1 | 144.0 | 170.3 | 15.4 |
| October | 187.4 | 165.8 | 172.3 | 3.8 |
| TOTAL | 127.2 | 548.2 | 756.1 | 27.5 |
| AVERAGE | 158.4 | 137.0 | 189.0 | 27.5 |

water consumption by 27.5 percent on average if tensiometers or similar field equipment is used to manage irrigation.

Ongoing research showed that causing water stress in coffee plants reduces the number of flowerings from the usual three patterns to only one in the same flowering season. This means improvements in the flavor of ground and roasted coffee. Considering the Brazilian practice of harvesting coffee in one single round, this is an important aspect. The quality of coffee decreases when there is a mixture of green, red, and black grains.

## 8.5 Future issues

The crucial aspects of environmentally sustainable food production in the Cerrado are threefold: the cost and availability of fertilizer; the preservation of water resources; and carbon sequestration. The accomplishment of using soil bacteria to fix nitrogen from the atmosphere after they are established inside the root nodules of some legumes, mainly soybeans, has already been mentioned. The problem is potassium. Brazil imports about 90 percent of its potassium (7.5 million mt/yr) for use in agriculture (US$2 billion per year).

A study led by CPAC (EMBRAPA Cerrados) and four other institutions is testing an alternative to this problem: rock powder. When specific rocks are powdered, they can release potassium in soils and become a source of low cost nutrients for plants. Rocks rich in biotite and phlogopite from the counties of Catalão and Rio Verde in GO and from Itabira, MG, have been intensively tested in greenhouses and experimental fields.

The research is coordinated by Dr. Eder de Souza Martins from CPAC (EMBRAPA Cerrados). According to him, some of the rocks in the study come from mining activities and, as with the use of lime to correct soil acidity, no chemical processing is involved. However, as the concentration of potassium ranges from 2 percent to 6 percent, transportation costs are a major issue regarding this technology. The timing of potassium release is another issue to be addressed.

As discussed above, some established agricultural practices that are environmentally unfriendly and have a negative impact on the climate, such as the flooding rice system (heavy methane emissions), livestock grazing on low-quality pasture (again, heavy methane emissions), and the conventional tillage system (often resulting in a high loss of soil organic matter and increased runoff and soil erosion), already have alternatives to reduce their impact on the environment.

Natural vegetation with a high priority for conservation of biodiversity can be addressed appropriately by federal and state laws. More efforts by decision makers and agricultural producers should be undertaken to preserve the water resources of the Cerrado. In the near future, water will be as valuable as gold and emeralds. Water resources zoning is one of the missing studies in Cerrado. People living in urban areas demand water for domestic consumption at levels of 300 liters/day/inhabitant. Crop irrigation by center-pivot demands water at levels of 1 liter/second/hectare. Is there enough water for such high-level consumption? Rainfall-runoff models from major basins of the Cerrado are incipient or non-existent. They are essential to determine the amount of water available for human consumption and agriculture. This amount can be easily estimated by the following equation:

$$R = I + ET + r - G + C \text{ (8.1)}$$

where $R$ = rainfall; $I$ = infiltration; $ET$ = evapotranspiration; $r$ = runoff; $G$ = contribution from ground water; and $C$ = water consumption.

If one proposes to estimate $R$, $I$, $ET$, $r$, and $G$ on a multi-temporal basis, then one can control the amount of water that will be available for consumption per hydrographic basin. This could be another public policy of the Brazilian federal government, like agricultural and sugarcane zoning.

A large-scale initiative to increase the reduction of carbon dioxide from the atmosphere by food production activity should also be implemented. One option is technical and financial assistance to farmers to recover degraded pasture. Data from scientific literature show that pastures with full biomass production capacity stoke more carbon in soils than degraded pastures. Therefore, surveys need to be conducted to know the total amount of carbon now stored in the soils of the Cerrado under cultivated pastures and what is the potential increase in storage if all pastures have a degradation level of zero. The major challenge here is how to extrapolate a few point-based field carbon measurements for the entire area occupied by planted pastures (54 million hectares). Multi-temporal remote sensing data will play a key role in this matter.

## Note

1. This figure is based on the EMBRAPA document. For details see Chapter 7 of this publication.

## References

Assad, E.D., F.R Marin, H.S. Pinto, and J. Zullo Jr. (2008) Zoneamento agrícola de riscos climáticos do Brasil: base teórica, pesquisa e desenvolvimento. *Informe Agropecuário*, 29, pp. 47–60.

Bastos Filho, G.B., D. Nakazone, G. Bruggemann, and H. Melo (2007) Uma avaliação do plantio direto no Brasil. *Revista Plantio Direto*, 101, pp. 14–17.

Bolliger, A., J. Magid, J.C.T Amado, F. Skora Neto, M.F.S. Ribeiro, A. Calegari, R. Ralisch, and R. Neergaard (2006) Taking stock of the Brazilian 'zero-till revolution': a review of landmark research and farmers' practice. *Advances in Agronomy*, 91, pp. 47–64.

Brannstrom, C., W. Jepson, A.M. Filippi, D. Redo, Z. Xu, and S. Ganesh (2008) Land change in the Brazilian savanna (Cerrado), 1986–2002: comparative analysis and implications for land-use policy. *Land-Use Policy*, 25, pp. 579–595.

Ferreira, L.G. and A.R. Huete (2004) Assessing the seasonal dynamics of the Brazilian Cerrado vegetation through the use of spectral vegetation indices. *International Journal of Remote Sensing*, 25(10), pp. 1837–1860.

IBGE (Brazilian Institute of Geography and Statistics) (2004) *Mapa de biomas do Brasil. Escala 1:5.000.000*. Available at: http://www.ibge.gov.br/home/geociencias/default_prod.shtm, date accessed May 8, 2013.

IBAMA (Brazilian Institute of Environment and Renewable Natural Resources) (2013) *Projeto de monitoramento do desmatamento nos biomas brasileiros por satélite – PMDBBS* (Brasília: IBAMA; Available at: http://siscom.ibama.gov.br/monitorabiomas/cerrado/index.htm, date accessed May 8, 2013).

Manzatto, C.V., E.D. Assad, J.F.M. Bacca, M.J. Zaroni, and S.E.M. Pereira (2009) *Zoneamento agroecológico da cana-de-açúcar* (Rio de Janeiro: EMBRAPA Solos, Documentos, 110).

MMA (Ministry of the Environment) (2013a). PPCerrado. Available at: http://www.mma.gov.br/florestas/controle-e-preven%C3%A7%C3%A3o-do-desmatamento/plano-de-a%C3%A7%C3%A3o-para-cerrado-%E2%80%93-ppcerrado, date accessed May 8, 2013.

MMA (Ministry of Environment) (2013b). Rural Environment Registry. Available at: http://www.car.gov.br/index.php/8-institucional/19-inicial-car, accessed May 8, 2013.

Ribeiro, J.F. and B.M.T. Walter (2008) 'Main phytophysiognomy of the tropical savanna biome' in S.M. Sano, S.F. Almeida, J.F. Ribeiro (eds), *Cerrado. Ecologia e Flora* (Planaltina: EMBRAPA Cerrados; Brasília: EMBRAPA Informação Tecnológica), pp. 151–199.

Sano, E.E. and F.S.C. Pinhati (2009) Espaço rural do oeste baiano: identificação de areas agrícolas sob sistema de plantio direto por meio de dados obtidos por camera digital e satellite CBERS-2 CCD. *Geografia*, 34(1), pp. 117–129.

Sano, S.M., S.P. Almeida, and J.F. Ribeiro, J.F. (2008) *Cerrado. Ecologia e Flora* (Planaltina: EMBRAPA Cerrados; Brasília: EMBRAPA Informação Tecnológica).

Sano, E.E., R. Rosa, J.L.S. Brito, and L.G. Ferreira (2010) Land cover mapping of the tropical savanna region in Brazil. *Environmental Monitoring and Assessment*, 166, pp. 113–124.

Trecenti, R. (2011) *Integração lavoura-pecuária-floresta* (Portal ILFP, Montes Claros: Universidade Federal de Minas Gerais. Available at: http://www.ilpf.com.br/artigos/integracao.pdf, accessed May 8, 2013.

# 9

# PRODECER: An Innovative International Cooperation Program

*Roberto Rodrigues*

## Introduction

As many people know, Japan and Brazil have a long agricultural relationship, from which they have both enjoyed benefits. The first cooperation between the two countries in the field of agriculture goes back about a century, right after the time when the first group of Japanese immigrants came to Brazil for farming. They brought with them what turned out to be two great contributions to Brazilian agriculture.

The first contribution was technology for growing vegetables and fruits. It led to the introduction of new horticulture products to the tables of Brazilian households, turning thier salads into more gorgeous dishes thanks to the variety of products to choose from. This contribution also meant that precious green belts were established around the big cities in Brazil and suitable conditions were developed for creating a transport and logistics hub for fruits and vegetables.

The second contribution was an associative work ethic. Japanese immigrants shared a teamwork spirit. This was their drive to single-handedly grow and supply vegetables and fruits over decades, resulting in leading Brazilian cooperativism and creating huge agricultural cooperatives, such as the Agricultural Cooperative of Cotia and the Agricultural Cooperative of South Brazil.

The Japan–Brazil relationship in the field of agriculture entered its second phase during the 1970s, centering on the development of Cerrado agriculture by the Japanese–Brazilian Cooperation Program for Cerrados Development (PRODECER). In need of expanding its food security, Japan collaborated with Brazil to develop the Cerrado, which until then had not been considered suitable for farming by farmers in Brazil. In fact, this vast area was not valued at all. There

220

was even a famous expression, "It is all right to take Cerrado land if it is offered for free or inherited." This explained how poor this land was for farming.

It was not surprising that the area had such a poor reputation. No one would want to live on barren land where nothing can grow without neutralizing the soil with a large amount of lime and fertilizer. PRODECER played a decisive role in the development of the Cerrado, where agriculture was made possible by the development of new technology, largely thanks to the Brazilian Agricultural Research Corporation (EMBRAPA) and the existence of the cooperative system.

Along with the technological innovation brought by EMBRAPA, PRODECER constituted an institutional innovation, being carried out through existing agricultural cooperatives such as the Agricultural Cooperative of Cotia and the Agricultural Cooperative of South Brazil. Without the participation of the cooperatives, Cerrado agriculture would probably not have succeeded.

Consequently, Cerrado agriculture, being a breakthrough practice and made possible by PRODECER, has brought enormous benefits to Japan and Brazil. In view of the historical achievements, PRODECER was an innovative approach at the time that Cerrado agriculture was established in Brazil and truly something that could be an institutional innovation. Along with the technological innovation brought by EMBRAPA, they both were essential parts of the development of Cerrado agriculture. How was PRODECER characterized, and how did it achieve Cerrado agriculture? Also, how did Japan and Brazil prepare, negotiate, and coordinate to accomplish the realization of Cerrado agriculture? This chapter addresses these questions. First, the background and preparatory efforts for the realization of PRODECER are summarized (Section 9.1). Second, the start-up process of PRODECER is explained, focusing on the problems that had to be overcome (Section 9.2). Third, the main features of PRODECER in terms of institutional innovation and its effectiveness for Cerrado agricultural development are analyzed (Section 9.3). Finally, concluding remarks are presented (Section 9.4).

## 9.1 Preparation of PRODECER

The beginnings of PRODECER can be traced to 1971, when the first mission of the Japanese National *Purchasing* Federation of Agricultural Cooperative Associations (ZENKOREN), headed by its president Makoto Mihashi, visited Brazil by invitation of Antonio José Rodrigues Filho, President of the Organization of Cooperatives of São Paulo State

(OCESP). The mission came to Brazil in order to study the possibilities of importing corn produced by Brazilian cooperatives.

In 1973, the United States President, Richard Nixon, announced an embargo on the export of grain, mainly soybeans, from the United States to Japan. Soybeans were one of the country's main sources of food and the decision caused panic. The agricultural stock-breeding cooperatives, which fed their birds and animals with American corn, soybeans, and bran, realized then how deeply dependent they were on a sole supplier. Therefore, the Ministry of Agriculture, Forestry, and Fisheries (MAFF) of Japan sent abroad technical missions in search of regions suitable for the production of grains on a large-scale.

After visiting various countries in East Asia, a Japanese government mission was sent to undertake research in Brazil. It was directed by Sakuei Matsumoto, chief of MAFF's Economic Cooperation Division, and including another four specialists, the mission arrived in 1973 in search of a region suitable for the promotion of large-scale production of the grains they were interested in.

After studying the potentially viable regions, with the help of technicians and specialists from the São Paulo and Minas Gerais Secretary of Agriculture and the federal government, after gathering plentiful research material from distinguished institutions, such as the Agricultural Institute of Campinas (IAC, Instituto Agronômico de Campinas) and technical reports from the IBEC Research Institute (Rockefeller Foundation), and after visiting the pioneering Alto Paranaiba Agricultural Development Program (PADAP) with farmers from the Cotia Agricultural Cooperative in the county of São Gotardo (in a partnership with the Minas Gerais state government), the mission focused its interest on the country's central Cerrado region.

The report – "Basic thoughts regarding cooperation for development of Brazilian agriculture" – triggered successive measures from the government for its adequate institutional framing between the two countries, based on the bilateral cooperation defined by Decree 69.089, of August 4, 1971, which outlined the Brazil–Japan Technical Cooperation Basic Agreement.

In the political area, the Joint Note between President Ernesto Geisel and Prime Minister Kakuei Tanaka, during his visit to Brazil on September 18, 1974:

> favorably accepted the possibility of future associations of Brazilian capital and Japanese private companies for the undertaking of agricultural enterprises in Brazil. Such enterprises, which will rely on the

majority participation of Brazilian capital, may be dedicated to the production, industrialization and commercialization of agricultural products that shall primarily respond to the needs of the internal Brazilian market and designate a part of the production to exports. Both governments shall consider the suitable support for such agricultural enterprises.

In February 1975, the Japan International Cooperation Agency (JICA) sent a technical mission headed by Tomomi Ashikaga, Director of the Japanese Research and Agricultural and Forestry Planning Department, to study the possibilities for the realization of the Brazil–Japan agricultural development cooperation enterprise.

It is important to note that in May of the same year, the Japan Business Federation (KEIDANREN), at that time headed by Toshio Doko, created the Committee for the Development of Japan–Brazil Agricultural Cooperation, headed by Tomisaburo Hirai, which contained representatives of 11 of the biggest Japanese trading companies and of the Industrial Bank of Japan (later replaced by the Long-Term Credit Bank). Representatives from the GAIMUSHO (Ministry of Foreign Affairs, MOFA), the MAFF, and JICA were present as advisors.

A month later, Tadashi Kuraishi, MAFF minister, was invited by the Minister of Agriculture Allyson Paulinelli to visit Brazil and received an *aide memoire* defining the general aims of the joint enterprise. The same text was presented to Prime Minister Takeo Fukuda when he visited Brazil in August 1975.

Next, in February 1976, a large group headed by Takashi Hisamune, Vice-President of JICA, held several long meetings with representatives of the Brazilian government, chaired by Ambassador Ítalo Zapa, chief of the Asia, Africa and Oceania Department of the Brazilian Foreign Affairs Ministry (known as Itamaraty in Portuguese), and with the technical coordination of the Brazilian Ministry of Agriculture, and with representatives from the private sector, in order to discuss the first version of the cooperation program. On this occasion, the program received its final denomination as Japanese–Brazilian Cooperation Program for Cerrados Development (PRODECER, in Portuguese), with the following stipulations:

(1) It was to be a pilot project of 50,000 ha;
(2) It was to involve the setting-up of an Agricultural Development Company (CDA), with shareholders from both countries' private sectors, through each country's respective investment companies, to provide support for agricultural production;

(3)  The CDA's capital composition should be equally divided;
(4)  The Japanese government's resources should be carried through JICA as a loan to the CDA;
(5)  The land would be acquired by the CDA;
(6)  Investment companies would be set up in Brazil and Japan.

In order to formalize the CDA, a holding company called BRASAGRO was established in Brazil under the coordination of the Ministry of Agriculture, but with shareholding participation of the private sector and of public and private banks, which would hold 51 percent of the ordinary shares, and, on the Japanese side, the Japan–Brazil Agricultural Development Corporation (JADECO), which, with JICA and KEIDANREN, would hold 49 percent. JADECO would have the shareholding participation of JICA, with 50 percent of the shares, and of 44 private companies, including 16 private Japanese banks, headed by the then Long-Term Credit Bank, with 43.75 percent of the shares; as well as three individuals: Toshio Doko, president of KEIDANREN, Takashi Hisamune, vice-president of JICA, and Tatsuzo Mizukami, Counseling Director of Mitsui & Co.

BRASAGRO had the following shareholding participation: Minas Gerais Development Bank (BDMG) with 25.5 percent; the Brazilian Economic Development Bank (BNDE, now BNDES) with 11 percent; Banco do Brasil (BB) with 10 percent; the National Bank for Cooperative Credit (BNCC) with 10 percent; ACESITA (Cia Aços Especiais de Itabira) with 4.9 percent; the Banco América do Sul with 2.5 percent; and other private banks, private companies, and more than 20 individuals holding the remaining 36.1 percent.

The CDA was named the Agricultural Promotion Company (CAMPO) and was legally formed in November 1978.

## 9.2   Start-up of PRODECER

It took five years to set up the PRODECER's executive structure, which covered, among other things, the kind of governmental support it would receive, the technological development and rural extension methods to be adopted, and the protection of Japanese resources against losses due to exchange rate variation, which were the problems that existed before the signing of the Record of Discussions. During the debates that took place in that period, PRODECER's financing system was created and the Loan Agreement and the Project Agreement were signed. This last document contributed greatly to the efficient execution of PRODECER by clearly stating the division of responsibilities between the Japanese and Brazilian governments, in their various levels.

During the aforementioned five-year period, approximately 30 missions from Japan visited Brazil, and about a dozen missions from Brazil visited Japan. Such missions were composed of technicians, specialists, businesspeople, politicians, ministers of state and even the President of Brazil and Japanese Prime Ministers.

One of the last missions, before the signing of the contracts, consisted of over 25 members, including Yohei Mimura, Vice-President of the Mitsubishi Corporation, Tatsuzo Mizukami, Counselor Director of the Mitsui Trading Co., and Kazuo Haruna, Vice-President of the Marubeni Corporation. When visiting in five planes São Gotardo, MG, at the heart of the Alto Paranaiba Agricultural Development Program (PADAP), they saw that production in the Cerrado was already a success, and that the main producers were *nisseis*,[1] sons and daughters of members of the Cotia Agricultural Cooperative, and decided to join the project.

However, in spite of the positive decision of these Japanese business-people, the bureaucracy of both countries necessitated a long period of negotiation. For example, the Japanese understood that the Brazilian Central Bank (BACEN) should be the holder of loans in yen from Japan, with the responsibility of shouldering the operation's risks of exchange rate variation, since it would be impossible to ask the farmers and cooperatives, who were the final borrowers according to the rural credit system in force in Brazil, to assume such a risk. Since at that time the BACEN could not be the holder of the Japanese loans, negotiations continued for a year. The solution adopted was for the National Treasury to guarantee the operation.

Another deadlock was the Japanese demand that Brazil guarantee the supply of soybeans, which was the object of PRODECER's establishment. Minister Paulinelli had to consult with President Ernesto Geisel on the matter, and he wisely argued that: if there is a substantial rise in the offer of soybeans in the market, thanks to this bilateral cooperation, and if there is no internal demand at that time, the international price of soybeans will be stabilized and Japan will be able to buy from where there is availability. Which is precisely what happened.

In Brazil, in addition to the government's internal bureaucratic procedures, the antagonistic positions inside the agricultural sector, since this was a pioneering program, the political clashes in the National Congress, and the radical position adopted by the Pastoral de Igreja (Social Organization led by the Catholic Church in Brazil) with the support of the media, would make a substantial 'romance' about the PRODECER.

Finally, in the spring of 1979, in the presence of Michio Watanabe, Minister of MAFF, PRODECER launched its first base in Paracatu, MG; an occasion which saw the minister in the driver's seat of a tractor.

Paracatu became the center of PRODECER's pilot project, with the participation of the Cotia Agricultural Cooperative, of Nikkei (Japanese–Brazilians) businesspeople led by Katsuzo Yamamoto and Fujio Tachibana (president of the Banco América do Sul), of businesspeople from Minas Gerais, and of Cotia-seinens immigrants (young immigrants invited by Cotia in 1955).

PRODECER was then implemented in the counties of Iraí de Minas, Coromandel, Paracatu, and Paracatu-Entre Ribeiros, all of which were in the state of MG, incorporating 70,000 ha of the Cerrado into the production process, an area which until then had been used only for grazing, with investments of US$50 million. In addition, two companies were established: the Company for the Agro-Industrial Development of the Cerrado, which incorporated 10,120 ha, with Banco America do Sul as the major participator; and the Curral do Fogo Agro-industria, with 5,380 ha.

The enthusiasm for the enterprise, and its success, attracted to the region other businesspeople and farmers from the south of the country (generally called *gaúchos*), which according to some studies, in addition to tripling the initial investment of PRODECER, allowed for the services sector in the city of Paracatu and neighboring regions to grow, raising incomes and lowering unemployment.

The historical basis represented in the presence and success of the Japanese immigrants in Brazil enabled the two governments to sign such an expansive agreement, which resulted in the PRODECER. Thanks to Japan's technical and financial cooperation, in the amount of US$562.9 million, Brazil became, from a producer of 5.8 million mt of soybeans in 1975, the second biggest producer in the world in 2005, with 60 million mt, and the biggest exporter of soybeans in the world.

The Irrigation Equipment Financing Program (PROFIR) was equally successful, with US$86 million of financing from the Japanese government. This program introduced modern technologies and equipment such as the center-pivot irrigation system, capable of irrigating more than 150 ha with a single self-propelled unit. This technology guaranteed the production of wheat, soybean, fruits, and other foods, minimizing the risks associated with a lack of rain, and even ensuring two harvests of grains annually.

The aforementioned programs boosted Brazil, which was able to become the biggest food producer in the world in the next decades. PRODECER was implemented in three phases, I, II, and III, and finished in March 2001. The total area of the Cerrado, centered in the Central-West region, is approximately 204 million ha (about 5.5 times the area of Japan). This large area was originally highly acidic and for a long time was considered non-productive, from agronomical point of view, mainly due to problems related to the existence of aluminum and the lack of essential minerals.

## 9.3   Main features of PRODECER as an institutional innovation

### (a) Program based on formal agreements between Brazil and Japan

Before the beginning of each phase of PRODECER, agreements were made between the two countries, and the implementation of the project was based on those guidelines. The base agreement was defined as "Establishment of the program's guidelines", and was composed of the record of discussions – Record of discussions (R/D) and those decisions defined the basic guidelines of both countries regarding PRODECER.

As a consequence of the base agreement, a document entitled "Establishing of the financing execution structure" was created. The Loan agreement (L/A) established, among other things, the Japanese side's financing, the location of resources by the Brazilian side, and the financing system of the producers and the cooperatives.

Also as a consequence of the base agreement, a document entitled 'Project agreement' (P/A), which established the scope of the Program, was defined. The P/A established the scope of the execution of the program, based on the study for the program's phases I, II, and III. The P/A is where the core of each phases' concept of development is, and clearly stated the responsibilities of the federal government, the state governments, the financial institutions, the cooperatives, and CAMPO, among others, in order to effectively conduct the program.

In phase I of PRODECER, a document entitled Basic Agreement (B/A) was also signed, but it was discarded from phase II.

### (b) Method for creating development complexes

PRODECER had the goal of creating development complexes in the new agricultural frontier region of the Cerrado and fundamentally envisioned the creation of medium-sized farms by farmers who did not possess land. Key points of the program included the following:

* The producers would acquire everything: agricultural equipment, houses, production equipment, etc.
* A considerable initial investment was necessary.
* A great part of the resources would be allocated through financing.
* In the process of occupying the land, there was growing concern about the environment.
* The improvement of the socio-economic infrastructure by the state and city governments was necessary.

### (c) Settlement method

As was mentioned at the beginning of this chapter, the cooperatives' leadership in Brazil and in Japan already had the idea in terms of partnerships between both countries in the complementarity of production and supply of food. Therefore, when BRASAGRO and JADECO were created, their cooperatives – respectively the UNICOOPJAPAN (ZENKOREN's trading company) and the National Bank for Cooperative Credit (BNCC, Banco Nacional de Crédito Cooperativo) – participated with a significant part of their social capital.

The selection of producers was conducted mainly by the cooperatives, which were selected by the Ministry of Agriculture and the CAMPO from the more structured ones in the country. The cooperatives supported their members in various areas, including the acquisition of land, the supply of materials and machinery, strategic planning, commercializing services, the storage and processing of production, and technical assistance.

The pioneering attitude and the accomplishments made by the Nikkei cooperatives, were important, especially the Cotia Agricultural Cooperative – Central Cooperative, at the time considered to be the biggest producers' cooperative in Latin America, with its experiences of PADAP in São Gotardo, MG, in the Settlement Project of Novo Mundo in Paracatu, MG, during PRODECER-I; and in the Settlement Project Ouro Verde in Barreiras – currently the county of Luís Eduardo Magalhães, BA during PRODECER-II. Another Nikkei cooperative, The Central Cooperative South Brazil, with headquarters in São Paulo, was the coordinator of the Settlement Project in Minas Gerais, in the city of Guarda-Mor.

These two Nikkei cooperatives were responsible for the settling of 154 producers' families, which represents 22 percent of the total number of settlers in PRODECER. In all, 16 cooperatives from MG, PR, SP, MT, and GO were in charge of implementing the settlements for the entire PRODECER.

## (d) EMBRAPA's technical cooperation program

To make agricultural production feasible in the Cerrado region, in which the land had a high acidity and low natural fertility, the development of specific agricultural technologies, the diffusion of techniques for the correction and handling of the soil, the selection of suitable crops and varieties, and the release of resources for the implementation of the projects were indispensable. In the case of PRODECER, the resources for this development were allocated simultaneously to the project's undertaking and with the cooperation of EMBRAPA.

## (e) Financing scheme

In the conception of PRODECER, because of its characteristics, the allocation of resources for financing the producers with low-interest rates was planned. Therefore, in the pilot projects financed by JICA, resources from the Investment and Financing Fund were handled directly by the Brazilian Central Bank. As for the financing conducted by JBIC (formerly the Overseas Economic Cooperation Fund, OECF), it was provided by the General Projects Financing Fund and was handled by BACEN through JADECO, all of it with very favorable conditions.

The Brazilian government took on the exchange rate risk that could eventually affect Japanese resources, ensuring the payment of interest and the devolution of the principal. Therefore, through this special financing scheme and availability of resources, the reduction of costs was realized.

## (f) Financing stages

The steps leading to the undertaking of the financing and its main points were:

- Financial resources
  - The necessary resources were allocated by both countries. The percentage was 50 percent each in phases I and II, and in phase III 60 percent from the Japanese side and 40 percent from the Brazilian side.
  - In the case of the pilot projects, 50 percent of the Japanese funds were from the Investment and Financing Fund of JICA and the rest from private banks. In the expansion project (full-fledged project), resources from JBIC's General Projects Financing Fund were used.

- The resources from the Brazilian side were allocated by the federal government, resources handling institutions (banks), and beneficiaries (producers, cooperatives, and agricultural companies).
- Selection of producers
  - Banks conducted their own analysis of the register of candidates.
  - Around 80 percent of producers were pre-selected by the cooperatives and by CAMPO based on the criteria previously established.
  - The remaining producers were selected by CAMPO, especially from farmers with agricultural science degrees and technicians already established in the region at the time of the implementation of the projects.
  - The main criteria for the selection of producers were:
    - Not possessing a rural property;
    - Brazilian nationality;
    - Experience in agricultural activity;
    - Availability to live in the property or in the city of the project;
    - Possessing their own resources.
- Payment forms
  - Loans to the borrowers were made by banks which received resources from the source.
  - CAMPO would periodically confirm the need of financing resources for the program's borrowers and would then send a request to the Ministry of Agriculture, which, after analysis, would formalize the request to National Treasury Secretariat (BACEN/STN).
  - Each transfer bank, in turn, would request the necessary amount from BACEN/STN.
  - On receipt of the official request from the Ministry of Agriculture, BACEN/STN would ask JICA, JADECO, and the Japanese banks to release the funds.
  - The Brazilian part would add its own part to the external resources from the ordinary federal budget, in order to answer the transfer bank's request, which would finance the producers and their respective cooperatives.
- Repassing banks
  - Phase I – Banco de Desenvolvimento de Minas Gerais (BDMG);.
  - Phase II – BDMG, Banco do Brasil, Banco Nacional de Crédito Cooperativo (BNCC), Banco do Estado de Mato Grosso (BEMAT), Banco de Desenvolvimento da Bahia (DESENBANCO).
  - Phase III – Banco do Nordeste (BNB) and Banco do Brasil. .
- Financing conditions

*Table 9.1*   Examples of financing conditions for producers

| | Type of credit | Time limit (years) | Grace period (years) | Period | Annual charge | Limit |
|---|---|---|---|---|---|---|
| PRODECER-I | Land | 20 | 6 | 1979/82 | 10% | 100% |
| | Investment | 12 | 6 | 1979/82 | 10% | 80–100% |
| | Expenses | 3 | 1 | 1979/82 | 10% | 100% |
| PRODECER-II | Investment | 15 | 6 | 1985/86 | ORTN + 3% | 80–100% |
| | Expenses | 3 | 1 | 1989/90 | IPC + 12% | 80–100% |

*Source:* Author.

All the agreements between the two countries had contractual terms that established the category of Special Program to PRODECER, since it was a National Project, influencing the conditions under which the foreign loans were taken, which were considered to be rather favorable.

The terms and conditions of any loan or credit that was conceded to any participant in the project under the terms of the Special Program, in the cases in which comparable loans or credits could be obtained through other existing agricultural credit programs in the Federative Republic of Brazil, will not be more onerous for the project's participants than the [...] terms and conditions applicable to the loans or credit available to other borrowers from the Federative Republic of Brazil, especially those located in the Cerrado region.

The financing conditions for producers through PRODECER, nevertheless, were subject to countless alterations over time, according to the scenario at that moment, especially from PRODECER-II.

### (g) Management of PRODECER

According to PRODECER's concept, two structures were fundamental for its execution, in addition to the final beneficiaries, especially for its producers: CAMPO, as coordinator of the execution of the project; and the cooperatives as the farmers' organizing mechanism, pursuing executive responsibility, in both the implementation and the consolidation of the projects.

Since its inception, CAMPO was headed by staff from MG, and had Japanese Vice-Presidents, appointed by JADECO – mainly retired public servants, bankers, and former multinational employees. It followed the guidelines of the Ministry of Agriculture, especially because in terms of revenue it depended on it for a percentage of the debt derived from loans given to PRODECER farmers.

*PRODECER-I (1979–1983)*

In the first phase of the program, three areas in the MG with good infrastructure and a satisfactory level of organization and technical assistance were selected. Sixty thousand ha were incorporated for the production of soybeans, corn, and coffee, in addition to rice, as a pioneering crop, after the cleaning process, considering its tolerance to the natural acidity of the Cerrado soil. The implementation plan included the creation of a farm under the direct administration of CAMPO, with an area of 110,000 ha and with the goal of producing good-quality seed, as well as to be a heritage for future expansions, and two agricultural companies with private investor participation.

The total cost of implementing this project was US$50 million, and it involved the settlement of 92 farmers' families.

In PRODECER-I, two methods of opening the agricultural frontier were adopted: (a) settlement; (b) business. In 1982, when the first Japan–Brazil joint evaluation took place, the settlement type was considered more adequate.

After the conclusion of this phase, from 1983, a settlement project was executed in the Santa Rosa region, Paracatu county – Project Entre Ribeiros I – under the same rules, using exclusively resources from the Brazilian side and an area of 10,000 ha, which was the property of CAMPO. This project benefited more than 41 families, and approximately US$17 million were invested.

*PRODECER-II (1985–1990)*

This phase took as its base the good performance achieved by PRODECER-I. The characteristic of PRODECER-II was the implementation of four Cerrado projects in two areas with different natural conditions: in MT, influenced by the Amazon region, and in BA, which has influences from the semi-arid Caatinga region. The goal was to develop technologies that were suited to the respective climatic conditions. The total area of the project was 65,000 ha, in which agricultural activities were undertaken that combined livestock, perennial crops, and cereals, such as soybeans and corn. The total cost of the projects was an estimated US$100 million and 165 families were settled. The main difference from the first phase was that in this case the cooperatives were responsible for the purchase and transfer of lands to the settlers, responsibilities that had been taken by CAMPO in the first phase.

In the same period, PRODECER Expansion was conducted, with resources from JBIC's financing for Projects in General, in 11 areas in MG, GO, and MS, regions that were considered to be able to realize the same results obtained in the areas covered by PRODECER-I. The area covered reached 140,000 ha, with 380 settled families at a total cost of US$275 million.

### PRODECER-III (1995–2001)

This phase was conducted in the counties of Pedro Afonso, TO, and Balsas, MA, regions of low latitude, located north of the projects previously conducted. The phase had the goal of consolidating rural administration technologies, with the introduction of irrigation and of new crops and varieties, under climatic conditions with small differences, in the insolation period all year long.

The area covered by the project was 880,000 ha, and 80 families were settled, at a total cost of US$137.9 million. The area of each property was 1,000 ha, which was double that of the projects implemented during phases I and II.

In addition to the introduction of irrigation equipment, another characteristic was the creation of a natural reserve with an area equivalent to 50 percent of the property, while in the previous phases little more than 20 percent had been allocated as a reserve. Also, most of this area was preserved as a collective reserved area.

In all three phases, Brazilian and Japanese technicians played an important role and contributed to the success of the program and without them the extraordinary advances would not have been reached. It is difficult to mention all of them, but as a representative of all these employees, the name of the agronomist Isidoro Yamanaka must be remembered. His contribution to Brazil–Japan relations was crucial and his documentation of events essential to the compilation of this chapter.

## 9.4  Concluding remarks: PRODECER's final evaluation

In July 2001, the governments of Brazil and Japan conducted an evaluation of PRODECER. (Ministry of Agriculture, Livestock and Supply (MAPA) and JICA, 2002) It concluded that: it had a distinguished role amongst the bi-national agricultural development programs in the Cerrado region and also outside of that region, and reached direct and indirect results, such as:

(a)  Contribution to the steady supply of food to the world;
(b)  Socio-economic improvements, thanks to the development of the country's inner region;

(c)   Development of agri-business and stimulus to regional development;
(d)   Diversification of countries exporting to Japan, and others.

Without any doubt, PRODECER had vast repercussions as a bilateral cooperation project, especially in Japan, because it was a program that was able to substantially raise the production of food, which was in the interest of Japanese population: they know the meaning of the hardship of hunger, and are also aware that their country, amongst the developed countries, has the least food self-sufficiency.

## Note

1.  Nisseis means here Japanese Brazilians whose parents are Japanese immigrants.

## Reference

Ministry of Agriculture, Livestock and Supply (MAPA) and JICA (2002) *Japan-Brazil Agricultural Development Cooperation Programs in the Cerrado Region of Brazil: Joint Evaluation Study: General Report* (Brasília and Tokyo: MAPA and JICA).

# 10
## Role of CAMPO in PRODECER: A Successful "Coordination" Model for Agricultural Development
*Emiliano Pereira Botelho*

### Introduction

The Japanese–Brazilian Cooperation Program for Cerrados Development (PRODECER), a joint venture between the Japanese and Brazilian governments, consisted of three main components: (1) research cooperation projects aiming to develop the necessary technologies for Cerrado agriculture; (2) funding cooperation projects for securing funds for producers to settle in the Cerrado region in order to start production activities; and (3) 'coordination' among many stakeholders taking part in the projects by combining (1) and (2) to achieve agricultural production in the Cerrado region in accordance with project objectives and time schedules.[1]

The Agricultural Promotion Company (CAMPO) has played a major role in (3), that is, 'coordination' work to guide relevant agencies and farmers to achieve agricultural production at the site where they came to work together. It is believed that CAMPO played a significant role in realizing the grand vision that Brazil and Japan had in mind during the 1970s, when outsiders had envisioned the Cerrado's agricultural development idea as a kind of 'unattainable dream.' In this chapter, Section 10.1 explains the background behind the establishment of CAMPO and its main features; Section 10.2 outlines the role of CAMPO in PRODECER; Section 10.3 discusses some of the most important aspects of CAMPO's coordination work with a focus on (1) 'technology dissemination,' (2) debt reduction negotiations, and (3) the outcome of 'system design' for the establishment of an agricultural cooperative at each project site; and Section 10.4 presents CAMPO's current business and its prospects for development.

## 10.1   Main features of CAMPO

PRODECER had a complex structure combining a number of relevant government and private agencies from Japan and Brazil. A general 'coordination' entity was needed to guide them to efficiently carry out projects in line with its strategy. CAMPO was founded in Belo Horizonte, capital city of the state of Minas Gerais (MG), to act as the coordinator of PRODECER's implementation activities. It was founded jointly by Japan and Brazil in November 1978 under the laws of the new Corporation Act. Brazil established the Brazilian Agricultural Participation Company (BRASAGRO) with capital of 1,002,000 cruzeiro, the Minas Gerais State Bank being the largest shareholder. Japan established the Japan–Brazil Agricultural Development Corporation (JADECO) with capital of 20 billion yen, consisting of 10 billion yen from the Japan International Cooperation Agency (JICA) and the other 10 billion yen from 48 private corporations. The Brazilian side held 51 percent of the shares and the Japanese side held the other 49 percent.

The work content was specified in detail in Article 2, 'Project Operation,' of the Project Agreement signed by the Central Bank of Brazil (BACEN), JICA, and CAMPO in September 1979. The five primary tasks, including 'coordination' work, expected to be accomplished by CAMPO in the project are listed as follows:

(1)   Research, planning, and coordination work relevant to the project;
(2)   Agricultural technology dissemination;[2]
(3)   Recruitment and selection of agricultural producers and acquisition of land for settlements;
(4)   Investment and guarantee of resources for agricultural producers;
(5)   Support for loan applications and technical advice to agricultural producers.

As seen above, CAMPO was established as a 'settlement company' for handling a series of projects in a comprehensive manner, starting with land selection for settlements and giving all the necessary technical support to the farmers and their families.

## 10.2   PRODECER and CAMPO

Settlement in PRODECER generally followed the steps listed below, and CAMPO was involved in every one of them: (1) land selection and acquisition for settlements; (2) development of a settlement plan (comprising roads, residential areas, agricultural land parceling, and the construction of educational and health care facilities and management offices); (3) application and registration of the settlement plan; (4) selection of settlers; (5) land sale to settlers; (6) agricultural management guidance and life support for settlers; (7) loan support for settlers; (8) support for organizing settlers; and (9) closing of the management office following independence of the settlement.

The Paracatu project site, during the first phase of PRODECER, provides an overview of CAMPO's work in project settlement. Paracatu is located in the northwest of at 47° W and 16° S, and has an area of 8,232 km². Paracatu has three different zones, with less rain at lower altitudes: a lowland (altitude ranging from 400 to 500 meters) described as a 'natural greenhouse' due to its high humidity; a central plateau (from 700 to 800 meters), where Paracatu's urban area is located; and a high plateau (about 1,000 meters), where the PRODECER site was located. The project site covered 39,000 ha, comprising the Mundo Novo ('New World') settlement of 23,000 ha and two cultivation companies of 16,000 ha.[3]

In 1977, prior to PRODECER's launch, Paracatu's population was 22,000, concentrated in the urban area and small villages along the river. PRODECER boosted the development of Paracatu and the county's population grew to 84,000 by 2010 – a four-fold population increase in 30 years, while a rural depopulation trend was observed in Brazil during the same period. Agriculture in Paracatu had been transformed from the level of 'gathering and subsistence agriculture' to a highly productive, modern, and sustainable system. This transformation was due to 'people,' 'agricultural technologies,' and 'funds' sent in from the outside. The cultivation area of five main crops – soybeans, corn, rice, feijão beans, and sugarcane – grew five-fold between 1980 and 2009, and the production of the main four grain crops showed remarkable growth, from 31,000 mt in 1980 to 360,000 mt in 2009. Large-scale, center-pivot irrigation agriculture was introduced and widely practiced (Paracatu county currently has 530 center-pivots in operation, which makes it the biggest user of this irrigation system in the country).

The economic development in Paracatu also influenced the educational field. There were no universities in Paracatu before PRODECER, and university students had to relocate mainly to Brasilia or Belo Horizonte. Today, there is one state university and four private universities in the county, providing 25 courses for 10,000 students in areas including medicine, agriculture, law, and architecture. Students come from all over the country, but most of them are from MG and Goiás (GO), and the Federal District (DF).

In terms of the work performed by CAMPO in Paracatu, it started with the selection of land suitable for settlement prior to the beginning of PRODECER. After searching, by analyzing aerial photographs, for a flat plateau of considerable size where mechanized agriculture was possible, CAMPO closely examined candidate locations to visually confirm landforms, rivers, and road conditions from the air by a specialized air company. On the ground, agronomists from CAMPO conducted site surveys and narrowed the candidate locations using data from vegetation and soil analysis. Land register checks and sales negotiations with major landowners were conducted through local real estate agents, and specialized lawyers at CAMPO headquarters carefully checked the paperwork related to the land, especially the names of the landowners, which tended to be incomplete or forged. Procedures for acquiring land for the Mundo Novo settlement were completed by February 1979, and in the following year, the settlement area was subdivided into lots. A section along the river with rich vegetation at the time had been designated as a public native vegetation reserve (*condomínio*), a protected area where no agricultural development would be carried out and which has been preserved to this day. All lots were planned so as to face roads, and if possible, be next to the river.

The settlement of farmers was conducted through an agricultural cooperative. The Agricultural Cooperative of Cotia (hereafter, Cotia), a Japanese-affiliated organization, and the largest of its kind at the time, was selected to be in charge, and 48 farming households moved in. When more than one household asked for the same lot, a draw was held. CAMPO took the initiative in all of the processes and handled them in a way that would be called a 'participatory style' today, i.e., in a democratic and fair manner. It was premised that the settlers would live on their lots so that they could not become absentee landowners; however, they were allowed to live in a nearby urban area for reasons such as sending children to school. As it turned out, many settlers stayed on the back of large trucks or in huts to oversee construction themselves while developing the farmland.

CAMPO and Cotia set up their offices in Paracatu to support the settlers. The farmland development was completed in September 1981, and the construction of 140 km of roads and 100 km of rural electrification was finished the following year. The total investment was approximately 4.1 billion yen, excluding the infrastructure cost covered by the state government, with approximately 3.8 billion yen from PRODECER loans and the rest from the private funds of the settlers and Cotia. Thus, the project cost 100 million yen per 400 ha farm or lot.

Steps 1 through 4 can be called 'preparation steps,' in which CAMPO strategically selected development posts with an eye toward future growth and managed general coordination so that cooperation from relevant agencies at the federal, state, and county level, as well as participating cooperatives, could be attracted and all available resources could be collected for the Paracatu region. CAMPO's local office was set up on the PRODECER site to serve as a frontline headquarters for general coordination, including not only support for settlers but also communication with local residents. CAMPO took a hands-on approach, holding meetings at which it provided detailed information in order to gain the support and cooperation of the local residents.

CAMPO also played a major role in transferring agricultural technologies. Originally, commercial agriculture was not practiced in the Cerrado region, and knowledge of agricultural technologies was hardly available in the project sites, let alone of the latest research findings. For this reason, CAMPO met with agencies related to agriculture to create and distribute a 'technical guidance manual.'

In the 'implementation stage' (Steps 5–9), CAMPO managed the coordination of participation and support by relevant agencies in line with a development schedule, and carried out the settlement project smoothly. Critically, it enhanced the organizational strength of the participating cooperatives.

After the last step (9) of the 'implementation stage,' there remained several challenges to be faced – in particular the handling of debt issues facing settlers and agricultural cooperatives, which had been caused by hyperinflation in mid-1980s. The Brazilian economy became chaotic in the second and third phases of PRODECER (1985–1993), when a high interest rate policy was put into effect, seriously affecting agricultural sectors. CAMPO had a leading role in overcoming this issue. Especially the coordination that CAMPO performed in Steps (6–8) significantly contributed to the success of PRODECER.

## 10.3   CAMPO's coordination work in PRODECER

This section will cover three important activities that should be especially mentioned: (a) technology dissemination to promote and validate the latest technologies released by research institutions among settlers; (b) negotiation with banks on loans for settlers and debt reduction negotiations conducted in cooperation with agricultural cooperatives after debt issues occurred during the time of hyperinflation and high interest rates; and (c) systems design for the establishment and growth of a new agricultural cooperative based in each settlement to support rebuilding following the failure of the existing agricultural cooperatives.

## (1) Technology dissemination by 'technical guidance manual'

In the first phase of the PRODECER project, CAMPO managed an agricultural experiment station in Paracatu and a farm in Coromandel. These two experimental areas played an important role in developing on-farm research in cooperation with private and public research institutions, such as the Cerrado Agricultural Research Center (CPAC), Brazilian Agricultural Research Corporation (EMBRAPA), and Minas Gerais State Agricultural Research Corporation (EPAMIG), as well as with specialists from Japan. Meanwhile, CAMPO was engaged in work to disseminate the agricultural technologies developed by these agencies among farmers in the settlements in cooperation with agricultural cooperatives.

The 'technical guidance manual' played a remarkable role in this technology transfer process. In PRODECER, research institutions that received funds and equipment on the basis of the agreement between Japan and Brazil made outstanding progress in Cerrado agriculture research. (CPAC was at the core of such institutions.) However, no matter how much progress was made in enabling development of agricultural technologies and opening the path to agricultural innovation in the Cerrado region, it was of no use unless farmers knew about the latest technologies and recognized their effectiveness in turning possibility into reality. The 'technical guidance manual' (Figure 10.1) served as the vital link between the research institutions and the farmers themselves.

The technical guidance manual was prepared for each crop (considering the main aspects of the natural environment, such as weather, soil, and landforms) in each settlement and conditions, including distribution and consumption trends. CAMPO compiled the manuals in cooperation with relevant research institutions after intense discussions taking into account farmers' own experiences and other local information, which could be obtained only by CAMPO through its close contact with the settlers.

In the second phase of PRODECER, following the success of the first phase, the number and area of settlements increased immediately, and Mato Grosso (MT), Bahia (BA), Mato Grosso do Sul (MS), and GO were included in addition to MG in projects, covering a broader area

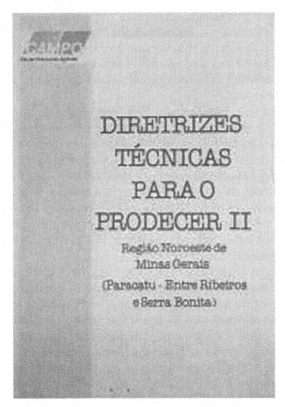

*Figure 10.1* 'Technical guidance manual' for the northwestern part of MG, including Paracatu, in PRODECER-II
*Source:* Authors (from the archive of CAMPO).

of the Cerrado region. Given that the natural environment, cultural conditions, and participating settlers and agricultural cooperatives became more diversified, it required considerable time and energy to compile a manual suitable for each settlement. Nevertheless, CAMPO also managed to collect information on settlements to take action by utilizing its mobility, comparable to that of a private company thanks to its semi-private nature.

**(2) Debt reduction negotiations with banks**
CAMPO moved its headquarters to Brasilia in April 1983, prior to the beginning of the second phase of PRODECER, due to the policy

of the federal government overseeing the program. As the project area expanded, close coordination with the Ministry of Agriculture, Livestock and Supply (MAPA) became inevitable, triggering the move. However, CAMPO was also in close contact and negotiations during PRODECER-II with banks functioning as loan agencies to lend project funds to producers.

CAMPO initially compiled a list of farmers eligible for loans. The final decision on the eligibility of the farmers was made by the lending bank, but, as a general rule, 80 percent of the farmers on the list were selected by the agricultural cooperatives and finally approved by CAMPO, and the remaining 20 percent were selected by CAMPO among local farmers. The primary criteria for selection were: not being a farmland owner; holding Brazilian citizenship; having some experience in farm production and management; being eager to participate in the projects; being ready to live on a farm or in a municipality; and having a certain amount of private funds.

Banks served as a lending agencies for the settlers. Funds were provided to the lending bank on the basis of a request from CAMPO, and originated from a program account for PRODECER held by the Central Bank of Brazil (BACEN). In other words, the lending institution (bank) processed a lending plan document prepared by CAMPO and officially requested BACEN (or the National Treasury of the Ministry of Finance from 1983) to provide financing. Upon receipt of the official request by the lending bank, BACEN asked JICA, JADECO, and banks on the Japanese side for financing. On the Brazilian side, the Japanese funds were combined with funds from the Brazilian national budget, to be loaned to the farmer and the agricultural cooperative to which the farmer belonged through the lending bank. For a summary of how the loans were handled, see Table 10.1.

CAMPO's 'coordination' work with the lending banks was a challenging issue. Particularly in the second phase of PRODECER, many lending banks participated in projects and responded differently from one another, posing severe difficulties in simultaneous negotiations. In the case of the debt problems, for example, farmers participating in PRODECER were granted debt reductions by state banks, such as the Bank of Development of MG (BDMG), whereas the Bank of Brazil (BB) was usually inflexible and frequently took tough measures, including the seizure of land and other property from farmers. Farmers had serious debt problems, which increased day by day due to hyperinflation during PRODECER-II between 1985 and 1993. For this reason,

*Table 10.1*   Banks lending funds to settling farmers in PRODECER

| First phase | Bank of Development of Minas Gerais State (BDMG) |
|---|---|
| Second phase | BDMG<br>Bank of Brazil (BB)<br>National Bank for Cooperative Credit (BNCC)<br>Bank of Mato Grosso State (BEMAT)<br>Bank of Development of Bahia State |
| Third phase | Bank of Northeastern Brazil (BNB)<br>Bank of Brazil (BB) |

*Source:* Japan–Brazil agricultural development cooperation programs in the Cerrado region of Brazil, joint evaluation study: General report (2002).

CAMPO could not easily withdraw from negotiations with the lending banks. Through the exercise of patience, it successfully reached a debt reduction agreement with many banks and contributed to reducing the farmers' burden.

According to the *Japan-Brazil agricultural development cooperaton programs in the Cerrado, joint evaluation study: General report* (2002), the percentage of farmers leaving pilot projects was 23 percent and those leaving full-scale projects was 27 percent in the second phase of PRODECER, lower than those of other settlement projects outside of PRODECER, which could be considered as proof that their burden was effectively reduced (Table 10.2). CAMPO's 'coordination' efforts with banks could therefore be considered as one of the factors behind the low rate of PRODECER farmers leaving farming.

Debt reduction was made possible by the trust CAMPO and settling farmers had built with the lending banks. CAMPO won their trust by incorporating the introduction of proven agricultural technologies in loan plans for the farmers to increase their productivity and at the same time prevent misuse of their funds. The loan plans for the farmers were first granted technical approval by CAMPO and then sent to lending banks to be reviewed. CAMPO used the aforementioned technical guidance manuals to assist with approval from a technical standpoint. Thus, it promoted agricultural technologies among farmers in accordance with the "technical guidance manuals", customized for conditions in each settlement. In the meantime, the company also built a system for loan funds to flow from lending banks on the condition of employing and complying with technologies that had been proven effectiveness and were expected to improve productivity. This strict 'project loan'

*Table 10.2* Number of settling farming households in PRODECER projects and rate of their leaving farming

| | At the beginning of settling (households) | Current (households) | Left farming (households) | Rate of leaving farming (%) |
|---|---|---|---|---|
| First phase Three PRODECER sites | 92 | 43 | 49 | 54 |
| (Entre Ribeiros I site) | (41) | (32) | (9) | (22) |
| Second phase Pilot project Four PRODECER sites | 165 | 145 | 20 | 23 |
| Second phase Full-fledged project 11 PRODECER sites | 380 | 278 | 102 | 27 |
| Third phase Two PRODECER sites | 80 | 80 | 0 | 0 |
| Total | 717 | 466 | 251 | 35 |

*Source:* Japan–Brazil agricultural development cooperation programs in the Cerrado region of Brazil, joint evaluation study: General report (2002).

system led state development banks to respond to debt reduction negotiations.

### (3) Agricultural cooperatives and system design

CAMPO had implemented PRODECER in cooperation with agricultural cooperatives from the beginning of the settlement. In particular, the cooperatives played a leading role in the selection of settlers. Also in terms of the above-mentioned technology dissemination and debt negotiation, it was essential for CAMPO to cooperate with the agricultural cooperatives, which were in a position to quickly find out the state of the settlements and settlers. Without their existence or cooperation with CAMPO, PRODECER would have struggled with an increased number of problems.

To begin with, the PRODECER settlements were mainly unused land owned by absentee landowners. There were also only a few isolated small-scale farmers. Therefore, agricultural cooperatives were initially non-existent in the region. This is why all the knowledge in the management of agricultural cooperatives, such as the Cotia, originally came from the experience of other regions and states. It was CAMPO that introduced and established a cooperative system to share the latest technologies in the Cerrado region and secure mutual support.

Furthermore, CAMPO was committed to the reorganization of the agricultural cooperatives. Hyperinflation in Brazil damaged their operation, leading to their dissolution one after another. Cotia was no exception, even though it was the largest of its kind in the nation and a contributor to the success of PRODECER by supporting the settlement. The cooperative defaulted on its debts in 1993 and announced its dissolution in 1994. Many agricultural cooperatives were forced to suspend their operations, and their property, including farm machinery, grain silos, and other real estate, was sold for debt repayment or seized by creditors. To continue farming, member farmers had to establish a new agricultural cooperative. CAMPO had dedicated a lot of effort to form a new agricultural cooperative in each settlement during these difficult times, using the technologies, experience, and trust it had obtained in the past (Table 10.3).

For instance, 40 families participated in PRODECER-III with 40,000 ha of land in Pedro Afonso, Tocantins (TO). The Cooperativa Mista de São João (COOPERSAN) was originally overseeing the settlement, but it was dissolved in the course of the project process due to its debt issues. Following the dissolution, farmers established the Agricultural Cooperative of Tocantins (COAPA) in 2000 and managed to continue the project, with CAMPO helping to design its system. CAMPO assumed responsibility for the 'coordination' of the technical and financial support necessary for the establishment of the new cooperative by negotiating with a bank, JICA, and a private consulting company. Thanks to its commitment to creating a new agricultural cooperative, the old one was replaced.

While financial issues caused many agricultural cooperatives to dissolve, CAMPO managed to complete the 'system design' of a better cooperative model, in which farmers would not be left to fend for themselves. It was important that the precious knowledge and experience that the old agricultural cooperatives, such as the Cotia, had accumulated were effectively incorporated into the new model.

Table 10.3  Settlement bases in PRODECER and agricultural cooperatives

| State | County | Project | Cooperative | Phase |
|---|---|---|---|---|
| MG | Coromandel | Coromandel | Associação dos Produtores Rurais de Coromandel | I |
|  | Paracatu | Mundo Novo | Coopernovo (Cooperativa Agrícola do Mundo Novo) |  |
|  |  | Entre Ribeiros I | Cooperativa Agropecuária Vale do Paracatu |  |
|  | Iraí de Minas | Iraí de Minas | Copamil (Cooperativa Agrícola Mista Iraí de Minas) | II |
| BA | Formosa do Rio Preto | Ouro Verde | Coproeste (Cooperativa Agrícola do Oeste da Bahia) |  |
|  |  | Brasil Central | Coaceral (Cooperativa Agrícola do Cerrado do Brasil Central) |  |
|  | Formoso | Piratinga | Coopertinga (Cooperativa Agropecuária da Região de Piratinga) |  |
|  | Buritis | Buritis | Coopago (Cooperativa Agropecuária do Planalto Goiano) |  |
|  | Bonfinópolis de Minas Unaí | Bonfinópolis | Coanor (Cooperativa Agropecuária do Noroeste Mineiro) |  |
| MG | Paracatu | Entre Ribeiros II | Coopervap (Cooperativa Agropecuária Vale do Paracatu) |  |
|  |  | Entre Ribeiros III |  |  |
|  |  | Entre Ribeiros IV |  |  |
|  | Guarda-Mor | Guarda-Mor | Cooperativa Agrícola do Oeste Mineiro |  |
| MS | Água Clara | Alvorada | Camas (Cooperativa Agrícola Mista de Alvorada do Sul) |  |
| GO | Ipameri / Campo Alegre de Goiás/Cristalina | Paineiras-Ipameri Cristalina | Coacer (Cooperativa Agropecuária do Cerrado) Cooperativa Agropecuária do Cerrado | III |
|  | Água Fria de Goiás/Alto Paraíso de Goiás/Niquelândia/São João d'Aliança | Buriti Alto | Copacen (Cooperativa dos Produtores Rurais do Planalto Central) |  |
| MT | Lucas do Rio Verde | Piúva- Lucas do Rio Verde | Cooperlucas (Cooperativa Agropecuária Lucas do Rio Verde Ltda.) |  |
|  | Tapurah | Ana Terra | Coopercana (Cooperativa Agropecuária Mista Canarana) |  |
| TO | Pedro Afonso | Pedro Afonso | Coapa (Cooperativa Agropecuária de Pedro Afonso) |  |
| MA | Balsas | Balsas | BATAVO-NE (Cooperativa Agropecuária Batavo Nordeste) |  |

Source: CAMPO.

## (4) Outcome of CAMPO's 'coordination' in PRODECER

In summary, CAMPO drove PRODECER to success, secured its position as a 'coordination' entity with relevant agencies to protect farmers' production activities and living, and made substantial contributions to: technology dissemination, utilizing the technical guidance manual to support loan procedures in order to minimize risks; debt reduction through negotiations on behalf of farmers; and system design for establishing and developing agricultural cooperatives for the protection of farmers and helping to increase their productivity.

CAMPO's success was made possible partly by its handling of public work with the speed usually seen in private companies. In its coordination work, CAMPO came into direct contact with farmers in 21 regions over seven states to understand their every need and finalized negotiations with the central/regional governments, lending banks, research institutions, and agricultural cooperatives. In a large-scale development project like PRODECER, governmental agencies are not expected to be flexible, and therefore take their time; circumstances could arise such that policy consistency may be in danger, as in the case of when the administration in the government changes accompanied by changes in its policies. In this regard, the accomplishment of CAMPO, a semi-private entity, in establishing and developing a public–private collaboration model in a practical manner, should provide rich insight for future agricultural development. Thanks to the success in PRODECER settlements, the autonomous and dynamic development of Cerrado agriculture was accelerated in the surrounding regions by the private sector.

## 10.4   Campo's activities after PRODECER

PRODECER ended on March 31, 2001 upon achievement of the planned goals. In its official evaluation at the end of the program, CAMPO was reported to have 'contributed to the smooth promotion of the PRODECER project.' After the completion of the program, CAMPO has continued to grow its business as a private company on the basis of the knowledge and experience acquired in PRODECER.[4]

Since the 1990s, CAMPO has diversified its business by expanding its presence in the agricultural field, using the specialized technologies developed in PRODECER as a basis, and has started to offer services with international markets in mind.

In 1991, CAMPO Centro de Tecnologia Agricola Ambiental was established in Paracatu. Aiming to increase crop productivity through plant nutrition improvement, it offered several services, including soil and plant material analyzes, water quality testing, and chemical fertilizer checking. The company laboratory is equipped with an official quality control system utilized nationwide and is acknowledged as one of the best laboratories in Brazil.

In 2002, CAMPO was restructured into a company comprising two segments: CAMPO Consultoria e Agronegócios, which offers project consulting and development services in related project fields, such as agricultural and agro-industry, settlement, irrigation and drainage, and the environment for companies and organizations, including international ones; and CAMPO Biotecnologia Vegetal, which was established in Paracatu in 1990 and conducts research and development into fruit, flowers, and tree seedlings, with a level of quality that is competitive in domestic and international markets and guaranteed to generate high productivity and commercial competitiveness. In 1998, for example, the company used tissue culture technology to produce sterile hybrid banana and apple seedlings resistant to serious pests. These were produced in a biofactory facility located at EMBRAPA's Cassava & Fruit National Center, in Cruz das Almas, BA.

By diversifying the business to the applied research and development of agricultural technologies, CAMPO continues to advance day by day, resulting in closer cooperation with research institutions performing high-level technical development, mainly in the technology transfer area.

With international markets in mind, it also runs foreign agricultural development consulting projects in an attempt to transfer its technologies to areas with similar vegetation to that of the Cerrado, i.e., tropical savannahs in South American regions and on the African continent.

The key to such technological transfer, as mentioned in this chapter, is 'coordination,' as practiced by CAMPO in PRODECER. There is no doubt that the research and development of innovative agricultural technologies and the securing of ample loan funds made an enormous contribution to PRODECER's success. However, the 'coordination' by CAMPO among many stakeholders, consisting of people and organizations from both the public and private sector, was also a key component in that success. With the help of CAMPO, PRODECER transformed the grain production structure of Brazil and helped to develop new agriculture in Brazil to counter threats to food security. The CAMPO of today is prepared to transfer its experiences from PRODECER to the world.

## Notes

1. PRODECER refers to the funding co-operation project listed as (2) in a strict sense but sometimes refers to the overall co-operation between Japan and Brazil for Cerrado development in general, the combination of (1), (2), and (3).
2. The Minas Gerais Technical Assistance and Rural Extension Corporation (EMATER-MG) was in charge of the technology dissemination activities for only the first term of the project.
3. Two methods were employed in the first phase of PRODECER. One was a "planned settlement project with farmer participation" and the other was "development by cultivation companies." It mainly aimed to compare their effectiveness. The former settlement method went quite well, and at the end of the project, the agricultural area had reached about 50,000 ha, two and a half times as large as the initial goal. On the other hand, the cultivation company method did not lead to effective results, mainly due to a lack of management skills among the participating companies. Therefore, only the planned settlement project method was employed from the second phase of PRODECER.
4. JADECO, a Japanese-funded company, completed its mission and went into liquidation proceedings in November 2007, resulting in the reduction of the percentage of the Japanese share in CAMPO from 49 percent to 9.96 percent, which is held by Japanese companies. BRASAGRO, an investment company on the Brazilian side, maintains its original 51 percent share.

## Bibliography

Diniz, Bernardo Palhares Campolina (2006) *O grande cerrado do Brasil central: geopolítica e economia* (Tese: doutrado/USP).

Gobbi, Wanderléia Aparecida de Oliveira (2004) Modernização agrícola no *cerrado* mineiro: os programas governamentais da década de 1970. *Caminhos de Geografia*, 5, pp. 130–149.

Inocêncio, Maria Erlan and Manoel Calaça (2010) Estado e território no Brasil: reflexões a partir da agricultura no *cerrado*. *Revista IDeAS*, 4(2), pp. 271–306.

MAPA and JICA (2002) *Japan–Brazil agricultural development cooperation programs in the Cerrado region of Brazil, joint evaluation: General report.*

Paiva, Denise Werneck de, Renato Fernando Amabile, Euclyde Minella, and Filipe Guerra Lopes (2006) Parceria interinstitucional público-privada na pesquisa agropecuária: o caso da cevada cervejeira. *Cadernos de Ciência & Tecnologia*, 23(2), pp. 235–251.

Pessôa, V.L.S. and Miguel César Sanchez (1989) Ação do estado e as transformações agrárias no *cerrado* das zoans de Paracatu e Alto Paranaíba (MG). *Boletim de Geografia Teorética*, 19, pp. 67–79.

Souza Júnior, Alcione de, Bernardo Palhares Campolina Diniz, Bernardo Pelegrini, Celso Norimitsu Mizumoto, Júlio César Augusto Sesma da Cruz, Paulo José San Martin, and Yoshisuke Ogura (eds) (2009) *O cerrado e o seu brilho* (São Paulo: Caramuru).

# Index

CPSIA information can be obtained
at www.ICGtesting.com
Printed in the USA
BVOW06*1106071117
499764BV00014B/377/P

9 781137 431349